The Architecture of "Society 5.0"

Hitachi-UTokyo Laboratory (H-UTokyo Lab.)
Editor

The Architecture of "Society 5.0"

Six Key Factors for a People-Centric
and Sustainable Smart City

 Springer

Editor
Hitachi-UTokyo Laboratory (H-UTokyo Lab.)
The University of Tokyo
Bunkyo-ku, Tokyo, Japan

ISBN 978-981-96-2928-2   ISBN 978-981-96-2929-9   (eBook)
https://doi.org/10.1007/978-981-96-2929-9

Translation from the Japanese language edition: "Society 5.0 Architecture" and The University of Tokyo
Joint Research Laboratory Copyright, © Hitachi and The University of Tokyo Joint Research 2023.
Published by Nikkei Publishing Inc.. All Rights Reserved.

This work was supported by the University of Tokyo and Hitachi, Ltd.

This Springer imprint is published by the registered company Springer Nature Singapore Pte Ltd.
The registered company address is: 152 Beach Road, #21-01/04 Gateway East, Singapore 189721,
Singapore

If disposing of this product, please recycle the paper.

Prologue: The Architecture as Blueprint to Society 5.0

Atsushi Deguchi

Society 5.0 as a Future Vision

In January 2016, the Government of Japan unveiled its 5th Science and Technology Basic Plan (Cabinet Office 2016, 2021a), which included a future vision titled Society 5.0, outlining Japan's vision for a technology-driven and people-centric future society. Since then, the government has worked to communicate this vision to domestic and international audiences. The vision is named "Society 5.0" because the envisaged society will be the "fifth" developmental stage of human society, following the hunter-gatherer (1.0), agricultural (2.0), industrial (3.0), and information (4.0) societies. As a people-centric society, Society 5.0 will balance economic goals with environmental and social concerns. The core technology for this purpose will seamlessly integrate cyber and physical spaces (or, as the 5th Science and Technology Basic Plan states, "the high degree of merging between cyberspace and physical space (real world)").

Hitachi-UTokyo Laboratory (H-UTokyo Lab)[1] released *Society 5.0: A People-Centric Super-smart Society* to explain the idea behind Society 5.0 and share research findings that have contributed to digitalization. The Japanese-language edition was published in October 2018 by Nikkei Publishing Inc. and has been widely read. During that time, interest in Japan's concept of the smart city, which involved energy-management initiatives for addressing the energy crisis Japan has faced since the 2011 Tohoku disaster, was growing. Therefore, we presented examples of smart grid projects and other existing energy-management initiatives and explained how Society 5.0 represents the next stage in these efforts. In the following years, several government strategies and projects were launched throughout Japan to realize Society 5.0, leading to a number of smart-city projects being implemented.

---

[1] Hitachi and the University of Tokyo established H-UTokyo Lab in June 2016 with a view to launching industry-academic collaboration for Society 5.0. H-UTokyo Lab is led by the Habitat Innovation Project team, which consists of members from Hitachi and the University of Tokyo. The project team authored *Society 5.0: A People-Centric Super-smart Society*.

An English edition of the text has been published in Springer Nature (H-UTokyo Lab 2020a), and a Mandarin edition has been published by China Machine Press (H-UTokyo Lab 2020b). These translations have helped broadcast the Society 5.0 vision across the world; thus, international audiences became familiarized with the concept, if only to a modest extent. We wish to thank all those involved in the compilation, translation, and publication of the text.

Social Challenges Brought About by the Digital and ICT Revolutions

Since 2016, social and technological progress has exceeded expectations. In 2016, Japan had already introduced a law regulating cryptocurrency, using "cryptoassets" as a blanket term. The digital economy started attracting increasing interest since 2020, with the rise and fall of Bitcoin.

At present, digital and information technology (IT) innovations pervade every corner of our lives. We have had to accept these innovations in a short space of time, and the innovations are still transforming our lives and living environments in breathtaking ways. History books tell us that the innovations and revolutions of the past benefited only a certain region and socioeconomic class. Many worry that the formidable benefits of digitization and IT—increased convenience, high profitability, etc.—will be enjoyed in a similar manner, by only a certain age group, a group of industries, and companies, leading to even wider regional, socioeconomic, and inter-business inequalities.

Such an outcome would be counterproductive to the UN's 17 Sustainable Development Goals (SDGs), adopted in 2015 as part of the 2030 Agenda for Sustainable Development. The SDGs have entered the everyday lexicon in Japan; in addition to being taught in schools, they are frequently mentioned on TV talk shows. The 2030 Agenda for Sustainable Development also included a promise that "no one will be left behind." If the world has committed to the SDGs and the "no one left behind" principle, the question becomes how digital innovation can advance in accordance with this sustainability agenda.

In September 2021, the Japanese government established the Digital Agency as part of a digital strategy. This digital strategy defines "digital transformation" in accordance with the conception in *Information Technology and the Good Life* (Stolterman and Fors 2004). In that work, Erik Stolterman and Anna Croon Fors (Umeå University, Sweden) suggested that digital transformation "can be understood as the changes that the digital technology causes or influences in all aspects of human life" and something that demands us to consider how "information technology can be in service of the good life"(Stolterman E & Croon Fors A 2004). This notion of digital transformation implies that digital innovations should proceed in a direction consistent with the vision proposed in Society 5.0.

However, contrary to this shared ideal, the ripples and benefits of digital technology are, as feared, reaching only specific regions, socioeconomic groups, and industries. Hurdles must be overcome to ensure that digital and IT technology is disseminated in the service of the good life, which is the true purpose of digital transformation and Society 5.0.

The Japanese government has committed to integrating digital technology with garden cities in a national strategy titled "Vision for a Digital Garden City Nation," which is already being implemented in locales across the country. In comparison, not much work has been done to establish an approach for a methodological framework for achieving a sustainable people-centered super-smart society. To address this lack of progress, H-UTokyo Lab has explored potential methodological frameworks. The purpose of this book is to disseminate these findings and contribute to smart-city initiatives that will help usher in Society 5.0.

Architecture Needed to Realize Society 5.0

In the early twentieth century, the garden city movement, an urban planning movement that gained worldwide prominence, began. The movement was the brainchild of Sir Ebenezer Howard (1850–1928), a British urban planner. In 1903, Howard founded a company named First Garden City to make his ideas on garden cities into a reality.

After the Industrial Revolution, societies grew increasingly industrialized (Society 3.0). In late-nineteenth- and early-twentieth-century Europe, many workers were living in urban areas full of poverty and squalor, and the specter of revolution haunted the continent. In his 1898 work, *To-Morrow: A Peaceful Path to Real Reform* (Howard 2003, 2007), Howard proposed his vision of garden cities that would combine the benefits of the town and the countryside. He also presented a corporate model in which a garden city would be function similar to a "large and well-appointed business." As the book's subtitle declared, the garden city movement was to be a "peaceful path," as opposed to a violent revolutionary one, to reform the problems that plagued cities after the Industrial Revolution. After the book's publication, Howard devoted the rest of his life to making garden cities a reality so that people could enjoy the benefits of towns and cities alike. Howard's vision and method were disseminated globally, leading to the creation of garden cities around the world.

Howard proposed both a vision and a method (which his corporate model amounted to) for the garden city, whereas Society 5.0 limits itself to being a conceptual vision; it simply points to a future that is driven by science and technology, as part of the government's basic science and technology strategy. It has never proposed any architecture. The onus is on local governments and businesses to establish methodological approaches for turning the vision into a reality.

In the present days, since the Covid-19 pandemic emerged in early 2020, we have had a further acceleration in the dissemination of digital technologies such as videoconferencing, creating palpable changes in our everyday lives; the rise of remote working is one example of such changes. This trend does involve some sacrifice—the social effects and the effects upon individuals are not all beneficial—but the accelerated rollout of infrastructure for remote working and learning has ushered in new working and learning, new choices in living spaces, new attitudes among consumers, and new ideas for organizing office spaces. The Web 3.0 era (a decentralized iteration of the Internet), metaverse spaces, and other emerging innovations are likely to both make life easier and lead to profound changes in people's values.

With the pandemic hastening the dissemination of digital technology and all the lifestyle changes that entailed, this time period could be an opportune juncture for proposing an architecture detailing how to leverage this digital infrastructure toward the construction of the people-centric sustainable society that Society 5.0 is supposed to be. Currently, however, no such architecture has been set out. All we have is a government publication titled Society 5.0 Reference Architecture (Cabinet Office 2020). To make Society 5.0 applicable to cities and communities, more detailed architecture must be created and shared with experts in smart cities and other relevant fields.

What Does "Architecture" Mean?

When Japanese people hear the word "architecture" (as the loanword *ākitekucha*), they will probably associate it first and foremost with its Japanese translation, *kenchiku*.[2] Although "architecture" can sometimes refer to physical structures, in a construction context, it has a different connotation. It refers to the art, techniques, and processes for building structures in a particular locale and time and to the architectural style of that locale and time (the meaning of architecture is discussed further in Chap. 1). Architecture in this context also connotes the idea of designing and building structures in a way that reconciles the abstract requirements of the client (living space requirements, work requirements, recreational needs, etc.) with the physical constraints of the locale, such as the local features and available materials (stone, timber, soil, etc.).

Thus, architecture refers to the art and method for building structures. Stated differently, it is a method, technique, and theoretical framework for creating a space that reconciles abstract requirements with physical constraints particular to a location. A physical building is the product of such architecture. The English word "architecture" is, therefore, an abstract and non-countable noun referring to the art of creating a space reconciling desired functions with local parameters.

---

[2] The Japanese term "*kenchiku*," initially adopted as a translation of "architecture," has undergone semantic expansion to encompass a broader spectrum of meanings, including those associated with "building" and "construction."

The term is also often used in contexts other than construction, such as in computer engineering, IT, and manufacturing (the auto industry and shipbuilding, for example). In such contexts, architecture serves as a higher-level concept encompassing the designs for the software and hardware. Thus, it is an important concept, though its precise meaning somewhat varies in different fields, each with an overarching structure, organizational setup, and distribution or arrangement of components. If you would like to learn more about the usage of the term in each field, please see the reference list.

This book presents a discussion on the architecture needed to make Society 5.0 a reality (as suggested by the title of this book), and Chaps. 1 and 3 present the role of such architecture in the key factors for creating a smart city that is people-centered and sustainable.

The Reference Architecture of Society 5.0 and Its Complementary Components

In 2011, the German government unveiled the concept of Industry 4.0, which focused on manufacturing, as exemplified by the smart factory.[3] By contrast, Society 5.0 is a broader vision, covering community-development planning with its concept of the smart city and covering manufacturing with its concept of the smart factory. The smart city is a key part of Society 5.0.

Developing a target community into a smart city in accordance with Society 5.0 will require applying the target community schemes for integrating cyber and physical spaces. The architectural approach (construction methods and implementation procedures) must align with local features and the values of the community. One of the purposes of this book is to provide a methodological framework to achieve this.

The government has provided some guidance here: a committee operating under the Cabinet Office has produced the *Society 5.0 Reference Architecture* (Cabinet Office 2020). Similarly, a reference architecture for the smart city has been released by the Cross-ministerial Strategic Innovation Promotion Program (which, likewise, operates under the Cabinet Office) (Cabinet Office 2023). However, in both cases, and especially in the case of the former, the architecture is very generic and emphasizes interoperability. Thus, the architecture requires refinements to ensure it can be applied in real-world communities for the creation and management of people-centered sustainable super cities. This process of refinement must involve integrating findings and best practices related to urban planning and resident-led community development.

---

[3] Industrie 4.0 is described (in Japanese) on page 143 of the following addendum, which describes initiatives in Europe and North America: https://www.soumu.go.jp/johotsusintokei/whitepaper/ja/h30/pdf/n3500000.pdf. Accessed March 7, 2024.

The addendum accompanies a 2018 white paper on information and communications published by the Ministry of Internal Affairs and Communications.

Chapter 2 presents a discussion on past public- and private-led efforts to construct smart cities. A guiding idea behind this book is that in the course of developing smart cities, resident perspectives must be incorporated into the architectural approach. Accordingly, this book suggests ways to refine the government's reference architecture for Society 5.0. In particular, we suggest ways for applying digital technology and data analytics in real-world urban communities to address problems and drive community development. Hence, this book is a handbook for Society 5.0 with a focus on architectural plus factors.

Key Factors for Preventing Smart City Initiatives from Being Short-Lived — Toward the Realization of Society 5.0

In March 2021, a time when the pandemic was in full swing, the Cabinet approved the government's sixth five-year strategy for science, technology, and innovation (STI Basic Plan) (Cabinet Office 2021a). Shaped by Japan's experience of the pandemic and with a focus on the post-pandemic transition, the Sixth STI Basic Plan also emphasized the construction of smart cities as a means to achieve the policy of ushering in a society that is "a sustainable and resilient society that ensures the safety and security of the people." The five-year STI Basic Plan is broken down into annual "integrated innovation strategies." The integrated innovation strategy for the year ended March 2022 (Cabinet Office 2021b), approved by Cabinet on June 2021, reiterated the need to "accelerate efforts" to make Society 5.0 a reality and identified smart cities as part of these efforts: (government translation) "In order to realize a society with a sustainable livelihood base, the government will seek to resolve issues faced by cities and regions, and aim to develop diverse, sustainable, and highly livable cities and regions [smart cities] nationwide that will proactively realize Society 5.0, which continues to create new values"; (retranslated: To support sustainable livelihoods, we will work to address the issues urban and rural communities face and support the national rollout of smart cities that are pro-diversity, sustainable, and livable habitable, and as such pave the way for Society 5.0, a society of endless value creation.).

In response to this national policy agenda, smart-city projects across Japan have established an overarching goal of making the Society 5.0 vision a reality. However, if the smart-city initiatives are to be implemented in a way that makes this vision a reality, they must be more than just a passing fad: the local governments and other actors must engage in the initiatives with a medium- and long-term focus.

Crucially, smart cities have a central role in the government's vision for a "digital garden-city nation." According to this vision, the goals of well-being and sustainability require the construction of smart cities that support people-centered communities and involve initiatives that are sustained over the medium or long term. The vision also identifies tasks that are critical to building smart cities that are people-centered and sustainable: Smart-city builders (local governments and other actors) must clarify what values to pursue and how to tailor smart-city designs to the features of the locality and community in question. Action plans must be developed to address these tasks.

H-UTokyo Lab has researched solutions to these issues. Drawing on these findings, this book clarifies the hurdles smart-city projects often encounter when pioneering applications for digital technology. It also proposes key factors that should be addressed regardless of the municipality or region in question to deliver a people-centered, sustainable, smart city.

Structure of This Book

As a follow-up to *Society 5.0: A People-Centric Super-smart Society*, this book is a compilation of the results achieved in the joint research on the architecture of Society 5.0 including the process and organizational infrastructure for building smart cities that embody the Society 5.0 vision, through industry-academia collaboration at H-UTokyo Lab. The book is designed to serve as a handbook for public officials in national and local government, for businesspeople, for academics, for those in the third sector, and for any other actor involved in this undertaking. We hope that the book will prove useful in understanding the current status of the smart-city agenda and its future path.

This book is divided into three broad sections. The first section is an introduction; it outlines the basic ideas behind Society 5.0 and the latest trends in the smart-city agenda. The section includes Chaps. 1 and 2. The first chapter clarifies our notion of architecture and the purpose of the book. The second chapter presents smart-city initiatives from around the world and a discussion on the policymaking context and approach to building smart cities. The chapter includes examples from Japan to illustrate the current status of the smart-city agenda in this country.

After these two chapters, the first section presents a discussion between the head of the University of Tokyo and the head of Hitachi. The discussion is titled "How to Achieve Society 5.0 and Deliver Wellbeing without Exceeding Planetary Boundaries." Since the Industrial Revolution, humankind has pursued relentless economic progress, as if Earth has unlimited resources and an unlimited capacity for cleansing itself. From now on, we must find ways to achieve well-being within planetary boundaries. In this discussion, the two leaders consider industry-academia collaboration that can contribute to a Society 5.0 that addresses global problems.

The second section of this book, which covers Chaps. 3, 4, 5, 6, 7, 8, and 9, introduces six factors critical to the success of efforts to build people-centered sustainable smart cities. These key factors are based on the findings of H-UTokyo Lab. Chapter 3 contextualizes the six key factors as augmentations to the government's reference architecture for Society 5.0. Chapter 4 outlines the hurdles that smart-city projects often run into and presents a discussion on how the key factors address these problems. Chapters 5, 6, 7, 8, 9, and 10 describe each key factor.

The six key factors are as follows: social acceptance, data governance, citizen participation, quality-of-life ratings in smart cities (QoL-Based Assessment), human resource development, and data ecosystems. Each factor represents something needed to enable a local government to build a smart city, address the local issues, and ensure that these efforts contribute toward a people-centered sustainable society. Based on the findings of H-UTokyo Lab and citing examples of practical knowledge and best practices, Chaps. 5, 6, 7, 8, 9, and 10 explain how each factor is crucial to ensuring that smart cities do not end up a passing fad.

Part III, covering Chaps. 11, 12, 13, and 14, describes three of H-UTokyo Lab's research projects as examples of practices for applying cyber–physical integration—a cornerstone of Society 5.0—in real-world urban environments. These projects provide insights for digitizing real-world data for optimal application in cyberspace and then feeding the outcomes back into the real world to address socio-environmental issues and enhance well-being. These insights provide an evidential basis for the key factors discussed in Part II (Chaps. 3, 4, 5, 6, 7, 8, and 10).

The final chapter of Part III, Chap. 14, explains how the six key factors, as factors pertaining to digital infrastructure, can be applied in real-world communities. The chapter also specifies a number of levels, envisaging a phased, stepwise development.

How to Use This Book

As a follow-up to *Society 5.0: A People-Centric Super-smart Society*, the Japanese edition of which H-UTokyo Lab published in October 2018, this book draws on the findings of research H-UTokyo Lab conducted subsequent to that date, and serves as a handbook or introductory reading. The 2018 book was intended to explain and disseminate Society 5.0 to a general audience (as opposed to an audience within a particular field) and then was largely unheard of, whereas the present book is more specialized by comparison.

This book is intended for two audience types: first, people representing public, private, or third-sector organizations engaging in the smart-city agenda; second, people whose expertise and engagement is in industries or fields related to the smart-city agenda. It contains information that is highly relevant to people involved in community-development planning (including urban planning, civil engineering, construction, and real estate); to healthcare experts, design experts, sociologists, economists, or researchers of public administration who are interested in cities and urban society; and to public employees working in national or local government. The book also offers referential material to undergraduate and graduate students of all disciplines.

As mentioned earlier, this book is broadly divided into three main parts. Each chapter is also written to stand on its own, so readers with limited time may choose to read only the chapters that align with their interests. For instance, after gaining an understanding of the basic concepts and the current trends in smart cities through Chaps. 1 and 2, the following reading paths may also be recommended:

- **For those interested in data utilization and citizen participation:**
  Chap. 5 (Social Acceptance), Chap. 7 (Citizen Participation), Chap. 11 (Public Dialogue in Data-Driven Urban Planning), and Chap. 14 (Key Factors as Infrastructure for a Digital Society)
- **For those interested in architecture and design:**
  Chap. 5 (Social Acceptance), Chap. 8 (Smart City QoL-Based Assessment), and Chap. 14 (Key Factors as Infrastructure for a Digital Society)
- **For those interested in civil engineering and infrastructure:**
  Chap. 7 (Citizen Participation), Chap. 8 (Smart City QoL-Based Assessment), Chap. 13, (City Infrastructure Management for Value Creation), and Chap.14 (Key Factors as Infrastructure for a Digital Society)
- **For those interested in health and medical fields:**
  Chap. 6 (Data Governance), Chap. 8 (Smart City QoL-Based Assessment), Chap. 12 (Smart Aging), and Chap. 14 (Key Factors as Infrastructure for a Digital Society)
- **For those interested in economics and public administration:**
  Chap. 6 (Data Governance), Chap. 7 (Citizen Participation), Chap. 10 (Data Ecosystem), Chap. 13 (City Infrastructure Management for Value Creation), and Chap. 14 (Key Factors as Infrastructure for a Digital Society)

The extent to which a chapter resonates with you may depend on your area of interest or expertise. Note that although much of the discourse on smart cities is focused on initiatives that use digital innovation, the focus of this book extends beyond technological aspects; it also emphasizes the overall architecture—the general structures and organizational designs that encompass digital initiatives among other things.

You should now have a clear idea about the purpose of this book and the general trends regarding Society 5.0 and smart cities, but before you proceed to the main body of the book, we have one more point to mention regarding urban environments and cyberspace: although cyberspace is often viewed as dichotomous with physical urban spaces, a key premise underlying this book is that the more cities embrace cyberspace, the more avenues they will find for urban development and renewal.

Cities consist of a complex array of urban "hardware"—roads, green spaces, sewerage. Such urban hardware should be built to last. These complex artifices must remain livable and be continually managed. A disposable city is incompatible with the need to live sustainably in harmony with nature. Individual structures can be replaced once they fall into a dilapidated state. The same cannot be said for a city. Cyberspace has a crucial role to play in ensuring a city's sustainability. If used effectively, it can make a city more livable and attractive. Smart cities, inasmuch as they align with the Society 5.0 vision of cyber–physical integration, help ensure that cities remain sustainable.

For smart cities, with their digitally powered innovations, this is only the beginning. We must ensure that the smart-city agenda never winds up being just a collection of isolated initiatives and a passing fad. Therefore, having public and local stakeholders with passion and commitment and the right systems and approaches is vital for new efforts to be made toward building a people-centered sustainable society. We hope this book serves such an end.

For a glossary of the nomenclature used in this book, see the glossary at the end of this book.

Department of Socio-Cultural Environmental   Atsushi Deguchi
Studies, Graduate School of Frontier Sciences   deguchi@edu.k.u-tokyo.ac.jp
The University of Tokyo
Tokyo, Japan

References

Cabinet Office (Council for Science, Technology and Innovation) (2016) *Dai 5 ki kagaku gijutsu kihon keikaku* [The 5th Science and Technology Basic Plan] (Released on January 22, 2016). https://www8.cao.go.jp/cstp/kihonkeikaku/index5.html. Accessed March 28, 2024.

Cabinet Office (2020) Smart City Reference Architecture White Paper, 1st edn. In: Cross-ministerial Strategic Innovation Promotion Program (SIP) Second Phase, Big-data and AI-enabled Cyberspace Technologies/Smart City Architecture Development/Smart City Architecture Design and Promotion of Related Verification Research (Released on March 31, 2020). https://www8.cao.go.jp/cstp/stmain/20200318siparchitecture.html. Accessed on March 7, 2024.

Cabinet Office (Council for Science, Technology and Innovation) (2021a) The 6th Science, Technology, and Innovation Basic Plan (Released on March 26, 2021). https://www8.cao.go.jp/cstp/english/sti_basic_plan.pdf. Accessed May 7, 2024.

Cabinet Office (Council for Science, Technology and Innovation) (2021b) Integrated innovation strategy 2021 (Released on June 18, 2021). https://www8.cao.go.jp/cstp/english/strategy_2021.pdf. Accessed May 7, 2024.

Cabinet Office (2023) *Sumāto Shiti: Rifarensu ākitekucha howaito pēpā* [Smart city: Reference architecture white paper], 2nd edn. In: Cross-ministerial Strategic Innovation Promotion Program (SIP) Second Phase, Big-data and AI-enabled Cyberspace Technologies/Smart City Architecture Development/Smart City Architecture Design and Promotion of Related Verification Research (Released on August 10, 2023). https://www8.cao.go.jp/cstp/stmain/20230810smartcity.html. Accessed March 7, 2024.

Howard E (2003, originally published in 1898) To-morrow: A peaceful path to reform, Routledge, London.

Howard E (2007, originally published in 1902) Garden cities of to-morrow, Routledge, London.

H-UTokyo Lab (2020a) *Society 5.0: A People-centric Super-smart Society*. Springer, Singapore. https://www.springer.com/gp/book/9789811529887. Accessed on March 7, 2024.

H-UTokyo Lab (2020b) 社会5.0:以人为中心的超级智能社会*[Society 5.0: A People-centric Super-smart Society]*. China Machine Press, Beijing, China.

Stolterman E & Croon Fors A (2004) Information technology and the good life. Information Systems Research: Relevant Theory and Informed Practice. 687–692.

Acknowledgments

Numerous individuals played integral roles in the composition and publication of this book. Although we cannot list every name here, we extend our heartfelt thanks to everyone who has offered their support and encouragement.

We extend our heartfelt gratitude to several colleagues from the University of Tokyo for their immense support: Makoto Gonokami, president of the National Research and Development Agency (RIKEN) and former president of the university; Teruo Fujii, the current president; Hiroaki Aihara, executive vice president; and Hiroaki Inagaki, former associate managing director and former director of the University Corporate Collaboration Department. From Hitachi, we are particularly thankful to Toshiaki Higashihara, executive chairman/representative executive officer: Keiji Kojima, CEO; Itaru Nishizawa, executive officer and CTO; and Norihiro Suzuki, chairman of Hitachi Research Institute and former executive officer and CTO of Hitachi.

Special thanks go to Shinobu Yoshimura, lab leader of the second phase lab and former vice president of the University of Tokyo, for his passionate and dedicated guidance and support during the second phase of H-UTokyo Lab's research projects, particularly with the Habitat Innovation Project.

We also appreciate the assistance of Miho Sugimoto, university research administrator at the University of Tokyo's Graduate School of Frontier Sciences, and other academics and university staff who played vital roles in the planning of this publication. We would like to thank Editage (www.editage.jp) for English language editing.

Additionally, we thank Shuichi Hirai of Nikkei Business Publications and Mei Han Lee (Editor, Spring Nature) for their significant contributions to the publishing process of this book.

December 2023
H-UTokyo Lab Habitat Innovation Project

Lab leader and project leader:
Atsushi Deguchi (The University of Tokyo)

Lab leader:
Hideyuki Matsuoka (Hitachi)

Project leader:
Tadashi Kaji (Hitachi)

Contents

Editor and Contributors

About the Editor

Hitachi-UTokyo Laboratory (H-UTokyo Lab.) Hitachi-UTokyo Laboratory (H-UTokyo Lab.) was established in June 2016 at the University of Tokyo through a partnership between the university and Hitachi. This initiative marks a departure from traditional industry-academia partnerships that focus solely on problem-solving. Instead, H-UTokyo Lab has pioneered the industry-academia collaboration model, leveraging the combined strengths of academia and industry to support the Government of Japan's Society 5.0 vision, which aims to realize a super-smart society. The lab's efforts include crafting and disseminating visions and developing innovative research and development programs tailored to meet the societal challenges essential for bringing this vision to fruition.

Contributors

Atsushi Deguchi Department of Socio-Cultural Environmental Studies, Graduate School of Frontier Sciences, The University of Tokyo, Tokyo, Japan

Teruo Fujii The University of Tokyo, Tokyo, Japan

Soichi Furuya AI Transformation Strategy, Digital Engineering Business Unit, Hitachi, Ltd., Tokyo, Japan

Toshiaki Higashihara Hitachi, Ltd., Tokyo, Japan

Chiaki Hirai Research & Development Group, Hitachi, Ltd., Tokyo, Japan

Yuki Igeta Graduate School of Frontier Sciences, The University of Tokyo, Tokyo, Japan

Katsuya Iijima The Institute of Gerontology/The Institute for Future Initiatives, The University of Tokyo, Tokyo, Japan

Tadashi Kaji Service Systems Innovation Center, Research & Development Group, Hitachi, Ltd., Tokyo, Japan

Kaori Karasawa Department of Social Psychology, Graduate School of Humanities and Sociology, The University of Tokyo, Tokyo, Japan

Takuya Kurita Graduate School of Frontier Sciences, The University of Tokyo, Tokyo, Japan

Izuru Makihara Political and Public Administration Systems, The Research Center for Advanced Science and Technology, The University of Tokyo, Tokyo, Japan

Hideyuki Matsuoka The Basic Research Center, Research & Development Group, Hitachi, Ltd., Tokyo, Japan

Tadashi Mima Smart Infrastructure Consulting Department, Hitachi Consulting, Tokyo, Japan

Graduate School of Media and Governance, Keio University, Kanagawa, Japan

Ken Naono The Digital Platform Innovation Center, Research & Development Group, Hitachi, Ltd., Tokyo, Japan

Shin Osaki Neighverse Inc., Tokyo, Japan

Sustainable Society Design Center, Graduate School of Frontier Sciences, The University of Tokyo, Tokyo, Japan

Tomoyo Sasao Faculty of Engineering, Reitaku University, Chiba, Japan

Yoshinori Sato The Design Center, Research & Development Group, Hitachi, Ltd., Tokyo, Japan

Kei Suzuki Environment & Energy Innovation Center, Research & Development Group, Hitachi, Ltd., Tokyo, Japan

Mitsuharu Tai Systems Innovation Center, Research & Development Group, Hitachi, Ltd., Tokyo, Japan

The University of Tokyo, Tokyo, Japan

Toshiya Watanabe Research and Innovation Office, The Institute of Science Tokyo, Tokyo, Japan

The University of Tokyo, Tokyo, Japan

Naoki Yoshimoto The Hitachi-UTokyo Laboratory, Research & Development Group, Hitachi, Ltd., Tokyo, Japan

Part I
Manifesting Society 5.0: The Evolution of Smart Cities

Chapter 1
What Defines the Architecture? An Approach to the Architecture of Society 5.0

Chiaki Hirai

Abstract In Chap. 1, we begin by explaining what "architecture" truly means, using Japanese castle architecture and urban design as examples. We demonstrate that architecture is not limited to individual structures, but rather represents a style that reflects the ideas and social structure of its time. From this foundation, we explore what modern urban architecture should entail, emphasizing the importance of physical and digital spaces. We argue for a people-centric approach to urban development that seamlessly integrates cyberspace with real-world spaces, aligning with the vision of Society 5.0. The chapter raises critical questions for realizing Society 5.0: "What are the essential factors for creating a people-centric smart city?", "How will these factors influence the architecture of Society 5.0?", and "What steps are necessary to achieve this architectural vision?" Addressing these questions is the primary goal of this book.

Keywords Society 5.0 · People-centric smart cities · Cyberspace and physical space · Architecture of Society 5.0 · Reference architecture

1.1 Architecture

Matsuyama City, located in Ehime Prefecture, flourished as a castle town during the Edo period (1603–1868). The still-intact *tenshu* (main keep) of Matsuyama Castle, situated at the city center's hilltop, offers a spectacular view of the town below. However, compared with the well-known Himeji Castle, also referred to as the "White Heron Castle" due to its enormity and its brilliant white color from the plaster, the appearance of Matsuyama Castle's keep may seem somewhat plain. Nevertheless, size and color are not the only distinguishing features between the two keeps; a key aspect is their structure—specifically, their architecture.

C. Hirai (✉)
Research & Development Group, Hitachi, Ltd., Tokyo, Japan
e-mail: chiaki.hirai.xj@hitachi.com

© The Author(s) 2025   3
Hitachi-UTokyo Laboratory (H-UTokyo Lab.), *The Architecture of "Society 5.0"*,
https://doi.org/10.1007/978-981-96-2929-9_1

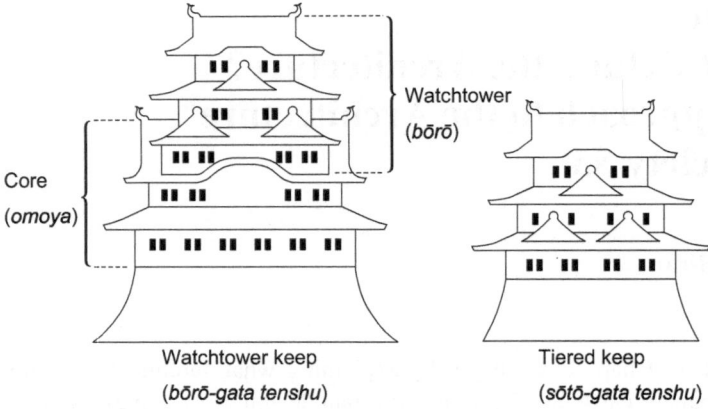

Core
(omoya)

Watchtower
(bōrō)

Watchtower keep
(bōrō-gata tenshu)

Tiered keep
(sōtō-gata tenshu)

Fig. 1.1 Architecture of *tenshu*

In the context of construction, the term "architecture" often refers to a particular architectural style. Architectural styles evolve over time. After a period has elapsed, similar architectural styles are retrospectively categorized and labeled as a specific "architecture" (for example, "Gothic architecture"). Thus, architecture serves as a conceptual model, with individual buildings being tangible instances of any of those models.

The Japanese castle keep originated from a structure combining a watchtower (*bōrō*) atop the core building (*omoya*), known as the "watchtower keep" (bōrō-gata tenshu) (see Fig. 1.1). Himeji Castle's keep exemplifies this architectural style, featuring a two-story core and a three-story watchtower (Hirai 1983). This composition gives Himeji Castle a complex and imposing impression. In contrast, Matsuyama Castle features a tiered keep (*sōtō-gata tenshu*), where the repetition of similarly designed tiers creates a simpler and plainer impression. Interestingly, initially built in the seventeenth century, Matsuyama Castle also boasted an imposing five-story "watchtower keep" style similar to that of Himeji Castle. According to legend, the architectural shift to a simpler tiered keep was made out of deference to the Shogunate (Doinaka 2002). The choice of architecture often reflects the demands of the era and expresses the ideas and philosophies of the time.

In addition to buildings, the concept of architecture is universal. For example, cars adhere to a common architecture. Once an individual learns to drive one car, they can virtually operate any car, regardless of its type or model, because all cars share the same underlying architecture. Similarly, computers exhibit standardized architecture; any mouse or monitor, regardless of brand, is compatible with any computer due to this standardization. The standardization of architecture encourages competition among suppliers, leading to reduced prices and fostering innovation. Thus, architecture benefits both users and manufacturers.

1.2   Urban Architecture

Cities also have distinct architectures. For instance, *Heijō-kyō* and *Heian-kyō* (present-day Nara and Kyoto, Japan's ancient capitals founded in the eighth century), both featured the same orthogonal grid layout. Suzaku Avenue ran from north to south, serving as the dividing line between the western and eastern halves of the cities. The western half was known as "*ukyo*," meaning "the right city," and the eastern half as "*sakyo*," or "the left city." These designations reflect the viewpoint of the emperor at the Palace, located at the northern end of Suzaku Avenue. Therefore, the architecture of these historic capitals embodied the concept of the emperor system—a state with an administrative and judicial system centered around the imperial court. As Japan transitioned to a samurai society, this urban design gradually evolved into the castle town architecture, which remains the fundamental urban layout in many of today's cities (Jinnai 1985).

During the Edo period, castle towns mirrored the feudal system, centered around feudal lords who resided in castles at the town center, with their high-ranking vassals inhabiting mansions surrounding the castle complex. Residential areas for townsfolk were positioned across a moat from the vassals' areas. In Tokyo, originally established as a castle town in the seventeenth century, this layout is still evident from the Imperial Palace to Marunouchi and Yaesu. Edo Castle was transformed into the Imperial Palace after the Edo period ended. In Marunouchi, the grand mansions once belonging to vassals have been replaced by towering office buildings. Across a JR line that once formed part of the castle's moat, the Yaesu area, now divided into small blocks, reflects its history as a residential area for common people. Similarly, in Matsuyama City, public administration buildings, such as the prefectural and municipal offices, are situated around the castle's outskirts. The thoroughfare, once lined with merchant houses during the Edo period, remains bustling with shoppers today. Walking around a historic castle town in Japan with an Edo-period map reveals that despite changes in the buildings, the historic urban architecture remains significantly unchanged and continues to dominate the scenery.

Urban architecture, likewise, reflects the ideology of the place and the time. Many European cities trace their origins to Roman times. Roman cities were surrounded by a rectangular perimeter of defensive walls. They consisted of an orthogonal grid, with principal streets (*decumanus maximus* and *cardo maximus*) intersecting near a central plaza known as the forum. The forums.

were places for the political discussions and debates that epitomized the Roman Republic. Eventually, statues of Roman emperors were erected there, a visible sign to the populace that Rome had transitioned from a republic to an empire. The layout of the Roman city was retained in Medieval Europe, but the thought of the age led to architectural changes. Roman-origin cities such as Milan, Vienna, and London switched to a layout that featured two city centers: the feudal lord's castle and a cathedral. The choices made in urban architecture reflect the mores of the place and time (thus manifesting philosophy and thought).

1.3 Tomorrow's Urban Architecture: Digital Architecture

What will be the architecture for the Japanese cities of tomorrow? This is a central question in this book. Japan's castle towns were founded on a feudal hierarchy, but the urban spaces of Society 5.0 must be founded on people-centric principles (see the introduction). What architecture will they require?

The cities of ancient and medieval times could be architecturally renewed with urban planning and government diktat; however, Japan's present-day urban spaces, where many people make a living, cannot be transformed quickly and expeditiously. One approach here is to focus on the use of underground infrastructure. Metro networks, sewer systems, electricity cables, gas lines, communication cables, and other underground infrastructure are vital lifelines for a city; without them functioning, city life grinds to a halt. There is something else that supports modern-day city life, though we barely register its existence most of the time: cyberspace.

Society 5.0 will be founded upon the seamless integration of virtual and physical spaces. Information in the physical world will be turned into digital data. The digitized data will be subjected to numerous calculations performed instantaneously in cyberspace. The results of the calculations will then be fed back into the real world to deliver services. This may sound like science fiction, but it is already happening, and we enjoy its benefits every day.

Consider, for example, taxi apps. A smartphone app connects you to a digital realm, with a digital map of the real world that provides real-time geospatial data collected from countless taxis. You can see at a glance which taxis are available for hire. These digitally represented taxis are, in a sense, conveying passengers around the digital city within the cyberspace.

As soon as you request a taxi, the app will scan the geospatial data to find a taxi near your location and calculate how long it will take for the taxi to reach you and how long it will take for it to deliver you to your destination. The calculations will be presented before your eyes in the physical world. Once you select a taxi, the app will call the taxi in cyberspace, and the real-world driver will come and fetch you. In this example, real-world information is digitized in real time, calculated in cyberspace, and the results are fed back into the real world, directing a taxi to the user's location. Food delivery services use the same system. If you recall how much of a godsend food delivery services proved during the pandemic, you get an idea of how cyberspace can serve as a flexible stand-in when something in the physical world is temporarily out of order. The more digital technologies develop, the greater the role of cyberspace becomes.

The architecture we need to discuss for Society 5.0 is not just physical urban structures. It means the information systems underlying cyberspace and the structures for society.

1.4 Reference Architecture for Society 5.0

The Japanese government (strictly speaking, the Cabinet Office) has released a reference architecture for Society 5.0 (Fig. 1.2). Before discussing this reference architecture, it is worth clarifying what a reference architecture is. A reference architecture is a model showing the recommended structures and designs for delivering technologies. Its recommendations are non-mandatory and non-prescriptive; the designer chooses whether or not to adopt the recommended structures and designs. It differs from "architecture" in the sense of retrospectively defined architectural styles for physical structures. It also differs from "architecture" in the sense of completely standardized architecture for computers and the like. It is a *reference* architecture, as in something "for your information." The reference architecture will be adopted if deemed sufficiently beneficial. The government has specified three benefits in its reference architecture for Society 5.0: First, it specifies processes for design and delivery, thus clarifying the procedure for developing Society 5.0. Second, it exhaustively covers all the required system components so that nothing

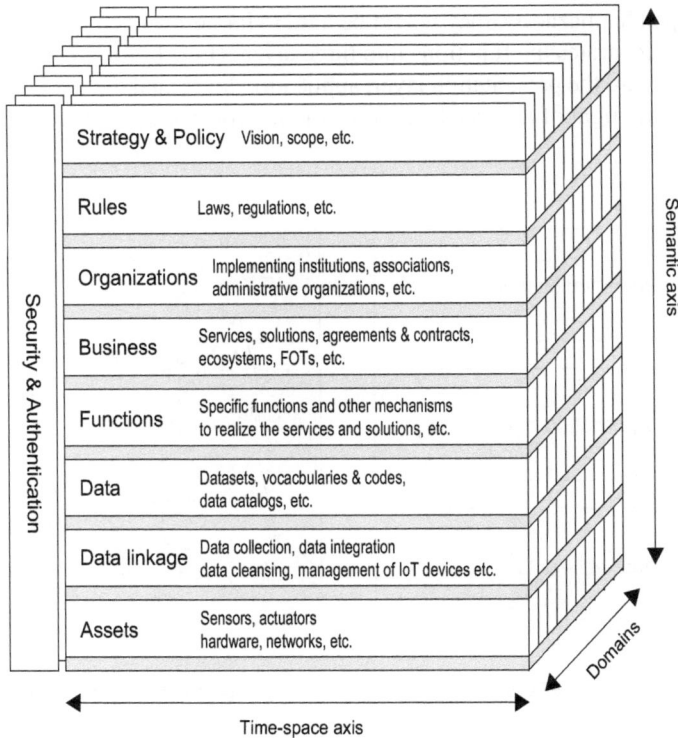

Fig. 1.2 Society 5.0 reference architecture (Cabinet Office 2020)

gets left undone. Third, as a universal architecture that can apply to disparate cities is provided, the outcomes are replicable and interoperable (Cabinet Office 2020).

However, a reference architecture is not the final word. When dealing with something as complex as a society, there is no guarantee that you will initially get the right architecture. It is a good idea to get feedback about possible tweaks, new perspectives, and extra detail. A reference architecture plays an important role in soliciting such feedback. In this book, we offer some suggestions for refining the reference architecture.

What, then, is the reference architecture for Society 5.0? What does architecture for society and cyberspace, as opposed to physical urban architecture, look like? In answering this question, a good place to start is a type of IT architecture that remains relatively simple and common: three-tier architecture.

1.5   Information System Architecture

Consider an application that shows the locations of available taxis on a map in real time.[1] The application is made from a vast number of functions. If these functions are developed haphazardly, they would be hard to harness and control. Three-tiered architecture separates the functions into three layers (Fig. 1.3a).

The layers are the presentation layer, application layer, and the data-access layer. They can be thought of as three boxes containing different assortments of parts and data items.

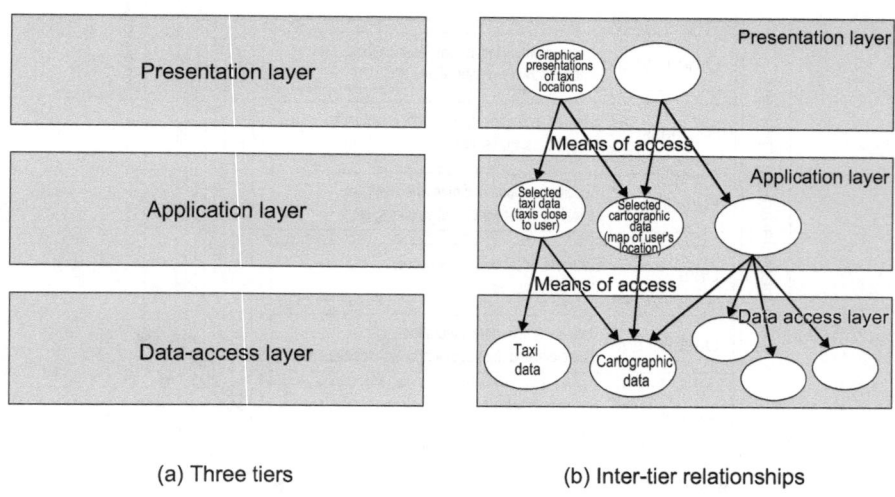

(a) Three tiers   (b) Inter-tier relationships

Fig. 1.3   Three-tiered architecture for IT systems

---

[1] The example discussed illustrates what is possible when functions are separated into the three-tiered architecture. Some taxi apps may differ from the example.

In the case of a ride-hailing app, the data-access layer (the bottom layer) consists of the data the app needs to access, including cartographic data and the geolocations of taxis that are for hire under the service. The presentation layer (the top layer) consists of interface functions, which in this example show the locations of available taxis on a map in real time. The application layer consists of functions that process and calculate the data and send the processed data to the presentation layer. In this example, these functions will select the available taxis and select the portion of the cartographic data giving a map of the user's surroundings. The hierarchy between these layers is vital. A layer can only access functions that are in the layer immediately below (Fig. 1.3b).

As this example illustrates, IT system architecture delineates boxes for separating the system's components and functions and defines the relationships between the boxes. What are the benefits of taking the time to define such architecture? First, the skills required of the developer vary considerably by layer. One set of specialists is required to design a user-friendly interface, another to design a slick algorithm, and another is necessary to ensure that the data are well organized and managed. Because any component belongs to any one of three layers, a specialist in charge of developing a particular component needs only to know one set of specialized knowledge. This arrangement enhances quality, development speed, and reliability.

The three-tiered architecture significantly shaped the IT system architectures that followed. It offers insights for understanding the reference architecture for Society 5.0 as well, as the next subchapter reveals.

1.6   Understanding the Reference Architecture for Society 5.0

Similar to the three-tiered architecture for IT systems, the reference architecture for Society 5.0 has a number of boxes, each containing a number of components. In Fig. 1.2, the "functions" box combines the presentation and application layers of the three-tiered architecture. In actual system design, the presentation layer is typically designed separately from the application layer. The "data" and "data-linkage" boxes correspond to the data-access layer. The "data" box contains data, while the "data-linkage" box contains an assortment of functions that mediate between the "data" box and the "assets" box directly beneath it. The "assets" box represents real-world hardware.

Let us summarize what we have learned about the bottom four levels of the reference architecture for Society 5.0. These layers include functions, data, and data linkage, which are adapted somewhat from the three-tiered architecture. They also include, at the bottom, assets, which represent hardware. This structure is a distinct feature of Society 5.0, with its seamless integration of cyber and physical spaces. In this structure, "assets" represent the real world, "functions" and "data" represent cyberspace, and "data linkage" integrates the two.

The top-four layers of the reference architecture are very different in meaning from the bottom four. Rather than representing software and hardware, they represent real-world organizations, institutions, and practices. The "strategy and policy" layer contains the purpose and vision statements for Society 5.0. The "rules" layer includes laws and regulations. The "organizations" layer contains the organizational actors who deliver the functions contained in the "business" layer directly beneath it. The relationship between layers in the top four layers is different from that in the bottom four layers, where a layer accesses functions in the layer directly below; in the case of these top-four layers, a layer governs the elements in the layer directly below. "Strategy and policy" governs "rules"; "rules" governs "organizations"; "organizations" governs "business." This hierarchy provides an order for planning structures and mechanisms for society. You start by devising a strategy and policies. You then set out rules that will enable the delivery of the policies (the rules could consist of new legislation or deregulation). With these rules in place, you then launch organizations that will engage in businesses (these businesses access functions in the "functions" layer below).

Now we focus on the part on the left: security and authentication. Running vertically down the whole figure, this component represents the things that adjoin every layer. It represents the things that protect each layer from the threats they will inevitably face—threats of misuse and destruction. Security and authentication can be likened to the whited walls of Himeji Castle, described at the beginning of the chapter. All tiers of the castle were plastered white in preparation for barrages of flaming arrows, for the defenders could never tell which part of the castle the flaming arrows would hit.

In summary, the reference architecture for Society 5.0 builds upon the three-tiered architecture of IT systems by adding four layers of real-world systems, with an "assets" layer beneath it, and provides security and authentication across all layers.

1.7 Balancing Economic Goals with Socioenvironmental Concerns

Some readers might find it puzzling to see a business layer in the reference architecture. This layer embodies the Society 5.0 principle of balancing economic growth with socioenvironmental concerns. This balancing often becomes challenging, as demonstrated during the pandemic-ravaged year of 2020, when the world faced a delicate equilibrium between sustaining economic activities and controlling infection rates. Enterprises within the "business" layer that address this balancing act. The layer encompasses public sectors, nonprofit enterprises, as well as social businesses that pursue profit growth and contribute to economic development, along with general private businesses.

1.8 Smart City as the Embodiment of a People-Centered Society

What, then, are the benefits of the reference architecture? The three benefits the government cited could apply to any reference architecture; they are not specific to this particular reference architecture. What sets this reference architecture apart is that its layers reflect the separate specialized skills needed for design and delivery like the three-tiered architecture for IT systems. This feature has the advantage of clarifying the prerequisite skills for each layer, making the work of specialists easier.

The question then arises: how do we integrate into this reference the other keyword and underlying principle for Society 5.0: people centrism? Is it enough that experts deliver Society 5.0 to laypeople and that laypeople enjoy the benefits thereof? This is a central question in this book.

Achieving people centrism requires understanding the values of residents. H-UTokyo Lab asserts that, through this understanding, the rather abstract notion of society can be translated into tangible community development, concept referred to as" Habitat Innovation" by H-UTokyo Lab (H-UTokyo Lab 2024). This book defines a smart city as something that professes and embodies the people-centric ethos of Society 5.0 and proceeds to identify key factors necessary for delivering a smart city.

In discussing architecture, we must never overlook the procedure for making the architecture a reality. To illustrate, let us revisit the realm of construction and consider an arch bridge. While having a blueprint that shows the finished appearance of the bridge is essential, the actual construction presents a different set of challenges. An arch becomes geometrically stable only upon completion; during construction, it remains unstable until the horizontal thrust is effectively distributed across the abutments at each end. Consequently, additional support structures are necessary while the construction is underway. Using ancient methods, builders would create a temporary soil mound to shape the arch with bricks, removing the mound only after the bridge's completion. Therefore, architecture involves more than just blueprints; it also encompasses the methods of construction.

Smart cities necessitate a construction method that adapts to changes in lifestyles and values. This approach should not entail the immediate establishment of a smart city in a single effort but rather offer a roadmap that charts a reasonable course toward completion. Throughout this process, it is crucial that residents feel confident in the progress of community development and never experience coercion. Community development must remain people-centered.

What are the key factors for building smart cities that are people centered as the Society 5.0 vision requires them to be? How are these factors reflected in the Society 5.0 architecture? What procedure is required to make the architecture a reality? This book sets out to answer these questions.

References

Cabinet Office (2020) Smart City Reference Architecture White Paper, 1ˢᵗ edn. In: Cross-ministerial Strategic Innovation Promotion Program (SIP) Second Phase, Big-data and AI-enabled Cyberspace Technologies /Smart City Architecture Development /Smart City Architecture Design and Promotion of Related Verification Research, p. 5. (Released on March 31, 2020). https://www8.cao.go.jp/cstp/stmain/20200318siparchitecture.html. Accessed on March 7, 2024.

Doinaka A (2002) *Matsuyama Jō no himitsu: Shiro to hanshu to jōka no kiso chishiki* [Matsuyama Castle's secret: Basic knowledge about the castle, lord, and castle town]. Atlas, Matsuyama, Ehime.

Hirai K (ed) (1983) *Momoyama kenchiku* [Momoyama buildings]. *Nihon no bijutsu* [Art of Japan], vol 200. Shibundo, Tokyo.

H-UTokyo Lab (2024) https://www.ht-lab.ducr.u-tokyo.ac.jp/en/summary-en/. Accessed on March 7, 2024.

Jinnai H (1985) *Tōkyō no kūkan jinrui-gaku* [Tokyo's spatial anthropology]. Chikuma Shobo, Tokyo.

Chapter 2
Trends in Smart Cities: Global and Japanese Perspectives

Tadashi Kaji, Takuya Kurita, Soichi Furuya, Yuki Igeta, and Atsushi Deguchi

Abstract This chapter first provides an overview of global trends in smart-city initiatives in Japan, Europe, the USA, and Asian countries before the Society 5.0 vision was proposed. It also describes the impact on smart-city initiatives during Covid-19. Subsequently, smart-city initiatives after Covid-19 are presented, and smart cities are the embodiment of Society 5.0. We discuss seven smart-city initiatives in Japan having a different set of conditions, including local features, the scope of initiatives, and actors. At the end of this chapter, we present the desired city for 2050 and its transformation scenario from the results of scenario analysis using an AI tool for policy recommendations.

Keywords Smart-city trends by region · Smart-city initiatives in Japan · Urban transformation scenario by AI · Super-city · Digital garden-city

T. Kaji (✉)
Service Systems Innovation Center, Research & Development Group, Hitachi, Ltd., Tokyo, Japan
e-mail: tadashi.kaji.ck@hitachi.com

T. Kurita · Y. Igeta
Graduate School of Frontier Sciences, The University of Tokyo, Tokyo, Japan
e-mail: takuya.kurita@edu.k.u-tokyo.ac.jp; igeta.yuki@edu.k.u-tokyo.ac.jp

S. Furuya
AI Transformation Strategy, Digital Engineering Business Unit, Hitachi, Ltd., Tokyo, Japan
e-mail: soichi.furuya.xz@hitachi.com

A. Deguchi
Department of Socio-Cultural Environmental Studies, Graduate School of Frontier Sciences, The University of Tokyo, Tokyo, Japan
e-mail: deguchi@edu.k.u-tokyo.ac.jp

Hitachi-UTokyo Laboratory (H-UTokyo Lab.), *The Architecture of "Society 5.0"*,
https://doi.org/10.1007/978-981-96-2929-9_2

In this chapter, we define a smart city as an entity that professes and embodies the people-centric ethos of Society 5.0. However, we recognize that smart-city initiatives were already underway before the Society 5.0 vision was unveiled. Smart cities differ depending on the requirements of the region and must evolve with time. Figure 2.1 presents a synopsis of smart-city trends, illustrating the objectives of such initiatives.

Year	Japan	EU
2007		■SET Plan : Strategic Energy Technology Plan
2008	☐Project for facilitating infrastructural measures in low-carbon-footprint urban development (MLIT, MOE) ☐Ecological urban development project (MLIT) ☐Eco-model cities (CO) Obihiro, Shimokawa, Iida, Yuzuhara, Minamata, etc.	
2009		■Directive 2009/28/EC of the European Parliament and of the Council of April 23, 2009 on the promotion of the use of energy from renewable sources ■EU climate and energy package Amsterdam Smart City (The Netherlands)
2010	☐Next-generation energy and social systems testbed (METI) Yokohama, Toyota, Keihanna (Kyoto), Kitakyushu	☐Europe 2020 Smart Santanderr (Spain)
2011	☐Future Cities (CO) Kashiwa-No-Ha (Kashiwa), Higashi Matsushima, etc. ☐Smart community vision proliferation (METI)	■Energy Efficiency Plan 2011 ★EU Smart Cities Information System ★Smart City Expo World Congress Smart City Barcelona (Spain)
2012	■Eco-City Act ☐Project to promote urban development, residential, and transport models that create, store, and save energy (MLIT) ☐Project to promote ICT-based urban development (MIAC)	★European Innovation Partnership for Smart Cities and Communities) Copenhagen Connecting (Denmark)
2013	★Council to promote ICT-based urban development (MIAC) ☐Project to promote models of resident-led carbon reduction planning (MOE) Smart City Aizuwakamatsu	
2014	Fujisawa Sustainable Smart Town	☐Horizon 2020 ■Digital Agenda for Europe 2020 Copenhagen Clean Cluster (Denmark)
2015	☐Project to promote ICT-based urban, human, and employment development (MIAC)	★Alliance for the Internet of Things Innovation (AIOTI) Intelligent Sustainable Paris (France) Smart City Berlin Germany)
2016	■5th Science and Technology Basic Plan ■Basic Act on the Advancement of Public and Private Sector Data Utilization ■Comprehensive Strategy on Science, Technology, and Innovation	
2017	☐Project to promote data-based smart cities (MIAC) Sapporo, Takamatsu, Kakogawa, etc.	■General Data Protection Regulation (GDPR)
2018	☐SDGs FutureCity (CO) Toyama, Toshima Ward, etc. ★Expert meeting on delivering the supercity vision (CO)	
2019	☐Smart-city model project (MLIT) Kashi-No-Ha (Kashiwa), Utsunomiya, Kozoji (Kasugai) ☐Smart-City Public–Private Partnership Platform (CO, METI, MOE, MLIT) ★Super-City Smart-City Forum 2019 (Osaka) (CO) ★Smart-City Institute	
		★G20 declaration: Data-free flow with trust/
		★Covid-19 emerges, becomes pandemic
2020	■Revision to Act on National Strategic Special Zones and the Act on Special Districts for Structural Reform (Super-City Act) ☐Applications for super-city status launched (CO) ★Digital Nippon 2020 (LDP) ★Smart-City Reference Architecture (CO)	■Next Generation EU ☐Sharing Cities Programme
2021	■6th Science, Technology and Innovation Basic Plan ★Smart-City Guidebook (CO, MOE, METI, MLIT)	
2022	☐Designation of super-city zones (CO) Osaka, Tsukuba ☐Designation of digital garden-city zones (CO) Chino, Kaga, Kibichūō ■Basic Policy for Digital Garden-City Nation	

Fig. 2.1 Synopsis of smart-city trends

General legend	■Legislation, state plan, etc. □Project ★Other ※ Smart city case	Abbreviations for Japanese public institutions	METI = Ministry of Economy, Trade and Industry CO = Cabinet Office MIAC = Ministry of Internal Affairs and Communicatinos MLIT = Ministry of Land, Infrastructure, Transport and Tourism MOE = Ministry of the Environment LDP = Liberal Democratic Party

North America	China	India
■American Recovery and Reinvestment Act of 2009 Dubuque 2.0 (Dubuque) Data SF (San Francisco)		
JUMP Smart Maui (Hawaii)	■12th five-year plan	
■Digital Government Strategy	■Announcement about designation of regions and cities as low-carbon model zones ■Announcement of urban planning for national smart-city model	
□Smart America Challenge Chicago Tech Plan	□90 sites designated as national smart-city model cities	
□Global City Teams Challenge ■Digital Accountability and Transparency Act		■Smart Cities Mission
□Smart Cities Initiatives □Smart City Challenge (USDOT)	★China smart-city expo	
Smart Cincy (Cincinnati)	■13th five-year plan ★MoU for Sino-Japanese cooperation in smart-city development City Brain (Hangzhou)	□Round 1 of Smart Cities Challenge (20 cities)
Smart Columbus Sidewalk Toronto (announcement of its plan)	Xiong'an New Area	□Round 2 of Smart Cities Challenge (27 cities)
	Smart Suzhou Shanghai Smart City	□Round 3 of Smart Cities Challenge (30 cities)
		□Round 4 of Smart Cities Challenge (10 cities)
★G20 Global Smart Cities Alliance founded (Oct)		
Sidewalk Labs announces it will leave Quay Side PJ QuaySide PJ redevt. plan	Hangzhou health code (Hangzhou)	■PM Gati Shakti Self-reliant India campaign
	■14th five-year plan	

Fig. 2.1 (continued)

2.1   Before Society 5.0

2.1.1   Japanese Smart Cities: Energy Efficiency

Japan's smart-city projects focused on improving energy efficiency for the city as a whole. In the Japanese literature, they are therefore known as "energy-management smart cities." The focus on energy management was intended to address the challenge of climate change, contribute to the low-carbon transition, and ensure a stable balance between power supply and demand following the bitter experience of the March 2011 disaster. The main examples of "energy-management smart cities" were the cities that participated in government projects delivered in 2010 and 2011, including a METI-led project for "Next-generation energy and social systems testbed (METI 2010)," a METI-led project for "Smart community vision proliferation," and a Cabinet Office-led project titled "Future Cities (FutureCity Initiative 2019)." One other "energy-management smart city" was Kashi-no-Ha (Fig. 2.2).

2.1.2   European and North American Smart Cities: Solving Residents' Problems

Around the same time, smart-city projects were also undertaken in Europe. In 2010, the European Council agreed upon a 10-year strategy called Europe 2020 (European Commission 2010), which included Horizon 2020 (European Commission 2014), a framework for funding research and development. Under Horizon 2020, the EU supported R&D and implementation projects related to smart cities. Driven by such technological development, European smart cities deployed sensor technologies with an emphasis on resident perspectives and solutions for local problems. In the Japanese literature, they are known as "sensor smart cities." Sensors would be deployed around the city to monitor real-world information and make the

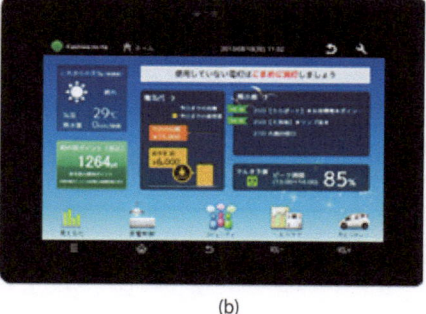

(a)   (b)

Fig. 2.2 Conceptual illustrations of Kashiwa-no-ha Smart City: Kashiwa-no-ha Smart Center (**a**) Kashiwa-no-ha HEMS tablet (**b**) (Hitachi et al. 2014)

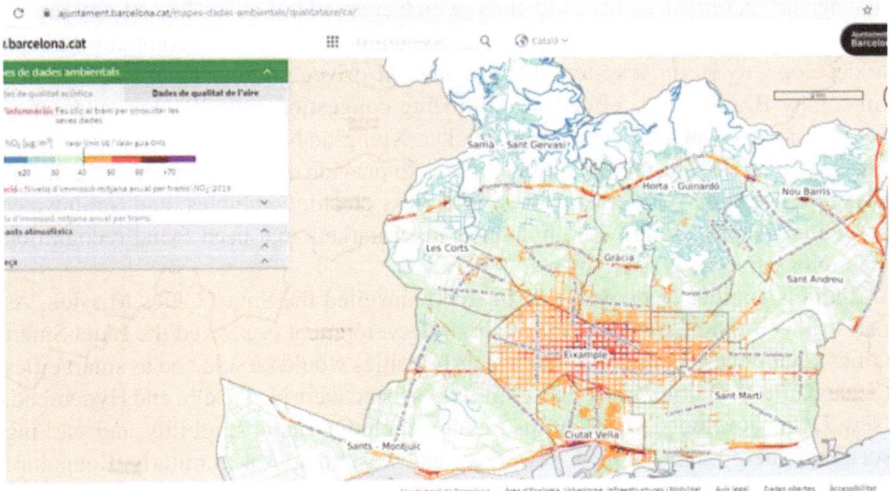

Fig. 2.3 Barcelonan sensor-based initiative for addressing air pollution problem (Info Barcelona, Barcelona City Council 2020)

information available as open-access data, enabling the visualization of local problems. The open data could then be used to develop practical solutions for the problems. One example was Barcelona, Spain. Barcelona was plagued with vehicular air and noise pollution. The city deployed sensors at intersections to monitor air pollution and noise levels. With the monitor data made available as open-access data, an application was developed to use the data: When readings in an intersection were high (indicating heavy pollution), the traffic signal patterns would be adjusted to minimize the traffic fumes around the intersection (Fig. 2.3).

2.1.3 Smart Cities in Continental Asia: Efficient Infrastructure

Many Asian countries other than Japan have pursued a national strategy for smart cities. A problem these countries faced was that urban infrastructure was failing to keep pace with the country's economic development, so smart cities represent an efficient means to supply the necessary infrastructure.

China is one example. In 2006, the government unveiled its 11th five-year plan. The plan included an agenda to build an energy-efficient circular economy, and the cities of Wuhan and Shenzhen were delegated as places to pilot initiatives to that end. In the years that followed, four big-name companies in China worked with local government to develop digital platforms for urban infrastructure, expediting the digital management of all functions required for regional administration (such as traffic, healthcare, urban management, environmental management). Thus, China's smart-city projects were characterized by the use of digital platforms to manage urban infrastructure efficiently, which enabled initiatives like energy saving and seamless mobility. The city of Hangzhou teamed up with tech company Alibaba to develop City Brain, a system that provides AI-driven monitoring of traffic in real time. City Brain proved effective in reducing congestion and allowing emergency vehicles to arrive at their destination quicker. Xiong'an New Area, established in the province of Heibei, used digital innovations to provide digitally driven social infrastructure, including driverless buses, driverless cleaning vehicles, and a rainwater recycling system, along with unmanned supermarkets that used facial recognition technology for cashless payments.

India is another example. In 2014, India unveiled the Smart Cities Mission. As part of this mission, the Ministry of Urban Development organized the India Smart Cities Challenge, a competition in which 100 cities would be selected as smart cities (Smart Cities Mission 2021). Two of the cities listed were New Delhi and Hyderabad. New Delhi developed smart infrastructure, including smart mobility and parking services, energy management, sewerage, and waste disposal. Similarly, Bengaluru launched a system of smart traffic management using automatic number-plate recognition. The two cities were selected in the smart-city challenge competition in recognition of how their initiatives demonstrated an understanding of local problems and emphasized financial sustainability. This example therefore illustrates that smart cities had progressed from a stage of pilot projects to a stage of full rollout.

2.2 Impact of Covid-19

2.2.1 Smart Cities Recognized as Digital Solutions for Preserving Life and the Economy

At the end of 2019, Covid-19 emerged and had a huge impact on smart cities. The pandemic led to lockdowns in cities. The biggest priority for smart cities became how to survive amid these circumstances.

During the pandemic, the smart cities used digital technology to prevail through lockdowns. Amid a shortage of facemasks, Taiwan's National Health Insurance Agency released CSV data showing the level of facemask stocks in pharmacies, with the data refreshed every 30 s. Using this data, the civic tech community and private sector developed map apps letting users see stock levels in nearby pharmacies and helping to ease the public disquiet over the mask shortage. In another example, Hangzhou used Alibaba's City Brain to develop an e-passport app called Health Code. Health Code checks the user's identity and then collates it with government-held big data to identify the user's health risk as one of three levels (red, yellow, or green). Other cities in Mainland China used robotics and drone technology to substitute manual work such as medical tests, health checks, goods transportation, rehabilitation, and disinfection, in order to reduce face-to-face contact and thereby mitigate the spread of infection.

2.2.2 Turning a Corner: Smart Cities Recognized as a Real-Estate Value Proposition

Some raised concerns about the lack of a business model for smart cities. They argued that the business case should be made for smart cities—namely, that by making transportation and other infrastructure more convenient, smart cities attract more people, raising property prices, and ultimately delivering profit growth.

However, this business model proved problematic when the pandemic prompted changes in people's lifestyles and mindsets; for example, it prompted people to avoid the three Cs (closed spaces, crowded places, and close-contact situations) and encouraged more remote working. Illustrating this business problem, Google affiliate Sidewalk Labs announced it would withdraw from Sidewalk Toronto, Toronto's project for a high-tech neighborhood. Sidewalk Labs cited "economic uncertainty" that had "set in around the world and in the Toronto real estate market (Sidewalk Toronto 2024)," revealing that the company had pinned hopes on a business model that would generate profit from real estate.

2.3 Smart Cities Get a Fresh Start

2.3.1 Super Cities

The Japanese government created the designation of "super city." Officially known as Super-City National Strategic Special Zones, super-smart cities are classified as "national strategic special zones" that pioneer initiatives for 2030 with community participation and guided by resident perspectives (Secretariat for Promotion of Regional Revitalization, Cabinet Office 2024a, b). Super cities were legislated for in September 2020 with an amendment to the National Strategic Special Zones Act. The government started soliciting super-city proposals in December 2020. By April 16, 2021, 31 municipalities had applied. In April 2022, two cities gained super-city status: Osaka and Tsukuba.

What sets super cities apart from smart cities is that super cities must conform to two explicitly stated requirements: They must listen to the community ("ascertain resident wishes") and must have an architect. "Ascertaining resident wishes" expresses the idea that a super city should involve citizen-led initiatives for addressing local problems, and is based on the premise that a super city requires public participation, with residents participating in and agreeing upon the future vision. Having an "architect" means specifying who is effectively responsible for the smart-city initiatives. It expresses the idea that a super city requires a team of creative and dynamic talent led by an architect.

Super-city applications were submitted during the pandemic. It is perhaps unsurprising, then, that many of the applications focused on digitally driven solutions for healthcare. In view of these applications, the government established a new category of strategic zone: the "digital-garden-health special zone." The government has grouped together zones into a cluster of local governments that engage together in concentrated efforts to develop digitally driven solutions for local healthcare problems, to which local digitalization projects and deregulation apply (Secretariat for Promotion of Regional Revitalization, Cabinet Office 2024a) (Secretariat for Promotion of Regional Revitalization, Cabinet Office 2024b). In April 2022, three municipalities gained this status: Kaga, Chino, and Kibichūō.

2.3.2 The "Digital Garden-city Nation"

The digital garden-city first surfaced as a vision for 2030 in Digital Nippon 2020, the Liberal Democratic Party's digitalization strategy, which was unveiled in June 2020 by a party committee for digitalization (Digitalization committee, LDP Policy Research Council 2020). The strategy envisaged Japan in 2030 as a "digital garden-city nation" that would be people-centered and benefit people in rural or peripheral regions (instead of just those in the main cities of Japan). It also stated that digitalization would level-up working practices, healthcare, and education in rural or peripheral regions. In November 2021, the government set up a council for delivering the vision (Council for the Realization of the Vision for a Digital Garden-City Nation 2022a, b). In June 2022, the Cabinet Office finalized a basic policy for the vision (Basic Policy for the Vision for a Digital Garden City Nation) (Council for the realization of the vision for a digital garden-city nation, Cabinet Office 2022a, b). The basic policy, translated from the Japanese, defines a digital garden-city nation as "a society that makes living and working everywhere in the country comfortable and convenient to live and work in." In a digital garden-city nation, the strategy continues; citizens can enjoy quality of life and live according to their lifestyle preferences and needs, regardless of where they are and regardless of age or gender. It is, in other words, "a society that creates new services for better working and living in rural or peripheral regions, is more sustainable, and supports better well-being, ensuring that digitalization benefits all citizens and service providers." This vision requires that digital technology be harnessed to address the socioeconomic problems in rural and peripheral regions, which are at the frontlines of problems such as depopulation and aging. Under the digital garden-city agenda, the government has funded numerous smart-city projects that harness digital technology to address local problems or make a local area more attractive. The government's grant program (the Digital Garden-City Nation Initiative Promotion Grant) delineates three categories of smart city (or digital garden city). In addition to funding pioneering projects that aim to establish a new model, the grant program funds other promising initiatives designed to be developed horizontally (Office for promotion of regional revitalization, Cabinet Office et al. 2022). In 2022, 27 organizations received funding under the program for a pioneering project, while 403 organizations received funding for a project to be horizontally developed.

Pioneering projects align with the idea of smart cities, as defined in the digital garden-city nation vision, in that they are required to monitor well-being metrics. The government is clearly prioritizing efforts to support better well-being under a people-centered paradigm and has signaled a policy of pursuing evidence-based solutions.

2.3.3 G20 Global Smart Cities Alliance

To create a digital garden city (or another version of a smart city), it is necessary to incorporate the best practices demonstrated in projects conducted in other regions. To that end, we need interoperable designs, but we also need to gain deeper insights into these other smart-city initiatives. With smart cities being built around the world, we have seen collaboration between cities that are located within a particular region of the world or that share common objectives. Japan has developed its own programs for inter-city collaboration. These include Smart-City Institute Japan (Smart City Institute Japan 2024)and the Ministry of Land, Infrastructure, Transport and Tourism's Smart-City Public–Private Partnership Platform (Smart City Public-Private Partnership Platform 2024).

At the 2019 G20 Osaka summit, hosted by Japan, the G20 nations affirmed that many of the hurdles cities face—privacy and security concerns, for example—are common to cities around the world, and that it is therefore necessary to have an alliance whereby cities around the world can learn from each other's best practices. This affirmation led to the creation, in June 2019, of the G20 Global Smart Cities Alliance (the full name is the G20 Global Smart Cities Alliance for Technology Governance) for the purpose of sharing best practices between cities regardless of region and objectives.

The alliance set out five common principles for cities: (1) openness and interoperability, (2) security and resilience, (3) privacy and transparency, (4) equity, inclusivity, and social impact, and (5) operational and financial sustainability. With this roadmap, the alliance examines the best practices of cities that have experienced similar problems and identifies model policies for successful smart cities (G20 Global Smart Cities Alliance for Technology Governance 2024).

2.4 Smart Cities as the Embodiments of Society 5.0

To summarize the previous section, smart cities started off as initiatives that harnessed digital and IT technology for certain objectives—for efficient energy management in the case of Japan, for citizen-led solutions in the case of Europe, and for more efficient urban development in the case of China and India (along with other Asian countries other than Japan). The pandemic then shifted the focus away from these initiatives, as cities around the world had to now focus on maximizing survival rates. As of 2023, the pandemic has abated and smart cities have gained a fresh start. However, the purpose of smart cities has changed somewhat from pre-pandemic times.

An analysis of recent smart-city initiatives suggests that smart cities can embody Society 5.0 in three ways: (1) by balancing well-being with resilience and sustainability, (2) being citizen-led, and (3) by harnessing digital technology for community support and mutual aid.

Supporting well-being is a way to further embody the Society 5.0 principle of people centrism. Japan's digital garden-city agenda involves setting quantitative metrics for well-being that will be digitally monitored as part of a systematic and ongoing effort to enable better well-being.

The citizen-led development, too, embodies the people-centric ethos. It denotes both citizen participation and citizen leadership. It also corresponds to one of the prerequisites for a super city, namely, listening to the community ("ascertain resident wishes"), which implies that residents actively participate in the process of delivering the vision. Just as the civic tech community played an active role in efforts to deal with the pandemic challenges, in post-pandemic times the public also has a key role to play in policymaking and delivery. The grand "architect" of the super city must encourage and coordinate such public participation.

As for community support and mutual aid, the digital garden-city agenda envisages intra-regional partnerships between citizens and between citizens and private enterprises, but it also envisages interregional and international sharing of best practices. Thus, as embodiments of Society 5.0, smart cities harness digital technology to build intra-regional, interregional, and international solidarity.

2.5 Reviewing Past Smart-City Initiatives in Japan

In Japan, main cities and peripheral cities have drawn from strategies set by the national government and pursued a smart-city format that aligns with the set of issues they face. Consequently, across the country as a whole, smart-city initiatives have covered a wide array of service areas, from disaster preparedness to transport services, logistics services, health and welfare services, and tourism. In this section, we discuss seven smart-city initiatives in Japan, each with a different set of conditions (including local features, scope of initiatives, actors) and some of these are still being piloted.

2.5.1 Kashiwa-no-ha Smart City (Kashiwa, Chiba)

The first of these cases is an example of a smart city developed on a tract of land set aside for redevelopment within a suburban region of the Greater Tokyo Area. The case in question is that of Kashiwa-no-ha, a zone established in the northern central area of Kashiwa City in Chiba Prefecture. August 2005 saw the opening of the

Tsukuba Express rail link linking Kashiwa to Tokyo. Following this, mixed-use development projects were launched in the vicinity of the local Tsukuba Express station, Kashiwa-no-ha Campus Station. These projects have been developed in tandem with an infrastructural development program that involved the rezoning of the northern central zone of northern Kashiwa. This program is conducted by Chiba Prefecture and runs from 2000 to 2029. It is designed to create a 273-hectare area with a projected population of 26 thousand. To the west of the station lies a tract of land that once was home to a US military base. It is now home to two satellite campuses (one belonging to the University of Tokyo and the other to Chiba University), the National Cancer Center Hospital East, and a green space (Kashiwa-no-ha Park). This tract also has a cluster of tech companies working on innovations in material development, AI, and life sciences.

Once the rezoning process delineated a zone around the station, the Urban Design Center of Kashiwa-no-ha (UDCK), a public–private-academic platform, started designing attractive urban spaces in a plaza and street west of the station and around a pond in Kashiwa-no-ha Park (no. 2 retention basin). In this way, it aims to establish a smart-city model that combines the creation of attractive public spaces and providing digitally driven public services, including driverless buses, AI-powered CCTV for security, and wellness support that draws on health data (Ishida & Kashiwagi 2019).

The first phase of this smart-city project began in the year ended March 2012, when the government designated Kashiwa-no-ha as an eco-friendly "future city" and as a "comprehensive special zone for regional revitalization." The second phase began in the year ended March 2020, when the Ministry of Land, Infrastructure, Transport and Tourism selected the area as a pioneer zone in its smart-city model project. The project continues to evolve.

During phase 1, Kashiwa-no-ha Smart City focused on energy management, as did other smart-city projects at the time, amid nationwide concern about energy security following the March 2011 disaster. Specifically, Kashiwa-no-ha Smart City focused on providing a type of energy-management system known in Japan as an "area energy management system" (AEMS). This system was launched in 2014, and applied to implement the smart city covering four zones around the station.

After the Society 5.0 vision was unveiled in January 2016, smart-city initiatives in Japan increasingly emphasized local problem-solving. Amid this new trend, Kashiwa-no-ha applied to the smart-city model project by the Ministry of Land, Infrastructure, Transport and Tourism. In 2019, it became one of the 15 pioneer zones selected by the ministry. The application had proposed a compact smart city around the station. Once the application was selected, Kashiwa City, Mitsui Fudosan, and UDCK led a consortium of private companies and research institutes (the Kashiwa-no-ha Smart-City Consortium), which in March 2020 published a strategy for delivering the project (Smart-City Action Plan) (Kashiwa local Govt website 2024).

The action plan set out four themes: mobility, energy, public spaces, and wellness. It also set out three strategies: build public–private data platforms, support open innovation with public–private academic platforms, and create interoperable services. The project seeks to drive community development by harnessing data related to people's life, wellness, and urban environments (Kashiwa smart city website 2024).

One of the issues in the zone was the need to improve mobility. There was some need for a bus service to cover the 2 kilometer distance between the station and facilities such as the campus and hospital. This problem fell under the "mobility" theme. In November 2019, the University of Tokyo teamed up with private-sector businesses to pilot driverless shuttle bus services linking the station to the university's Kashiwa campus. Four round-trip services were launched (one of which was for inspection purposes) with level 2 automated driving. As the next goal, the Kashiwa-no-ha area aims to introduce level 4 autonomous driving (full automation under specific conditions) supported by digital technology infrastructure in road spaces where autonomous and general vehicles coexist.

Under the first strategy (build public–private data platforms), Kashiwa-no-ha has provided a data-linkage platform called Dot to Dot, and wellness-related portal site called Smart Life Pass Kashiwa-no-ha (UDCK Town Management website 2024). Dot to Dot provides a safe environment for sharing the user's personal data among service providers. Smart Life Pass Kashiwa-no-ha uses Dot to Dot to provide health advice tailored to the user's health risks along with a menu for reserving or accessing local healthcare amenities and services. In these ways, both platforms contribute to wellness in Kashiwa-no-ha. Under the second strategy (support open innovation with public–private-academic platforms), Kashiwa-no-ha created a living lab called "studio for shared community development," which serves as a platform for residents and other stakeholders to participate in the process of developing the smart city. The living lab (which is discussed in greater detail in Chap. 7) hosts workshops for communicating local issues and needs and for generating ideas and prototypes for cocreating the smart city. In this way, the living lab serves as a launch pad for participatory projects (UDCK website 2024).

With the consortium increasing its membership year by year, even during the pandemic, Kashiwa-no-ha Smart City has the makings of a Japan-originated international model for a successful smart city.

2.5.2   Dai-Maru-Yū Smart City (Chiyoda, Tokyo)

The next case exemplifies as a smart-city project in business districts of central Tokyo. The project covers three such districts: Otemachi, Maru-no-Uchi, and Yurakucho, collectively known as Dai-Maru-Yū. The Dai-Maru-Yū zone is situated over a 120-hectare area in Chiyoda City (a city-level "special ward" of Tokyo). A prime business district of Japan, the zone encompasses 28 rail lines and 13 stations, with an employed population of 280 thousand and 4300 workplaces. Dai-Maru-Yū is also a shopping district as a result of strategic initiatives that began in the late 1980s. In the 2010s, Dai-Maru-Yū became known for pioneering initiatives in area-wide management and became a model for public–private initiatives between local

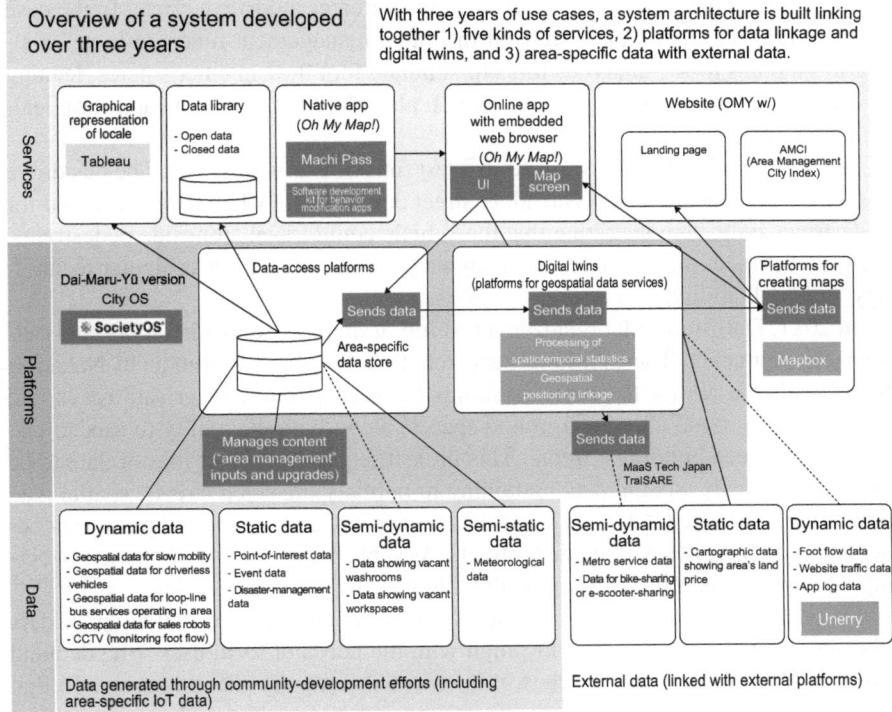

Fig. 2.4 Overview of the system for Dai-Maru-Yū (Dai-Maru-Yū smart city Website 2024)

firms and the Tokyo Metropolitan Government. The year 2019 saw the start of a full-fledged smart-city initiative here, which is designed to raise the area's value.

Dai-Maru-Yū currently uses a system that encompasses a wide array of datasets and services (see Fig. 2.4). The data are wide-ranging and granular in quantity and quality; they include dynamic data related to mobility services in the zone (such as sensor data for monitoring foot flows and driverless bus locations), static data related to events and disaster-management efforts, and semi-dynamic data such as information on vacant washrooms and workspaces. All these types of data are publicized as a zone dashboard, graphically representing what is happening in the zone.

All the urban data are analyzed and processed, along with external information (including metro service data and meteorological data), on a platform called Society OS (on a version of the platform specific to Dai-Maru-Yū). The processed data can then be used by apps to deliver services. Shown below are examples of pilot projects for delivering services that illustrate the digital transformation of "area management" (a Japanese-English buzzword that usually means public–private community-development efforts in a particular urban district) for Dai-Maru-Yū. In this way, real-world information is digitized and stored. The stored data are made available for analytics that can add value to the information. The added value is then fed back into the real world, improving services and enhancing the "area management" community building. This cyber–physical loop creates a positive circle. In the case of Dai-Maru-Yū, this look is dubbed the "area management redesign loop." Dai-Maru-Yū plans to use analytics tools to simulate foot flow in emergencies, leading to better disaster management in the area. It also plans to develop an app for a real-time tour map.

In 2020, Dai-Maru-Yū organized two test runs of a driverless bus. The bus drove at 6 km/h along Marunouchi Nakadori Street during hours when it was reserved for pedestrians only, demonstrating that the vehicle could travel alongside pedestrians. This pilot project offers interesting insights for balancing smart vehicular technology with walkability.

In 2021, Dai-Maru-Yū organized a test run of delivery robots in indoor and outdoor environments. The aim is to have robots totter along Marunouchi Nakadori Street with the ability to cross thresholds between public and private roads and between overground and underground spaces, along with the ability to ascend and descend between buildings' floors. This project requires several types of data to be linked organically: spatial data pertaining to publicly managed streets, spatial data pertaining to private land, and spatial data pertaining to elevators and passages connecting indoor spaces with outdoor streets. As such, the project illustrates a pioneering attempt to link urban 3D data with building information modeling.

One concern about smart-city projects is that the ideas the technology providers (the supply side) discover may misalign with the needs of local users (the demand side). To prevent such mismatches, it is essential to clarify the objectives. To that end, Dai-Maru-Yū has set key performance indicators (KPIs) that reflect local human needs related to creativity, comfort, and efficiency.

The longer term is also considered. On the assumption that the project will develop in an agile manner whenever new initiatives are tried between now and 2040, the project includes an extensive range of data and options and a robust organizational structure to enable various kinds of pioneering initiatives to be undertaken.

2.5.3 Smart-City Takeshiba (Minato, Tokyo)

We now explore the case of Takeshiba, a smart-city zone in Tokyo's Minato City. Takeshiba is an example of a project-oriented smart city. Adjoining Tokyo Bay, Takeshiba features a waterfront and port. As with Dai-Maru-Yū, Takeshiba is subject to public–private "area management": The area is managed by a council representing the private sector and local government, and operational matters are managed by a public-interest association. Takeshiba has a core facility called Tokyo Portcity Takeshiba. Over a thousand sensors have been deployed here to collect foot flow data and other urban data. The data is linked with a 3D city modeling program called "Plateau" and a building information modeling program, enabling simulations to be run for community-development initiatives (Fig. 2.5). A number of pilot projects have been conducted already. One example involved a robot taking food from a convenience store and delivering it to people in a building. Another involved

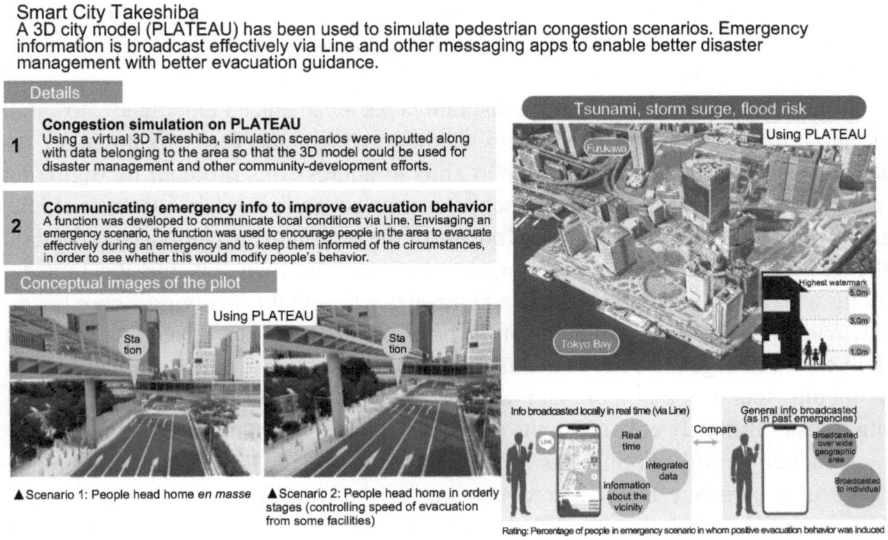

Smart City Takeshiba
A 3D city model (PLATEAU) has been used to simulate pedestrian congestion scenarios. Emergency
information is broadcast effectively via Line and other messaging apps to enable better disaster
management with better evacuation guidance.

Details

1 **Congestion simulation on PLATEAU**
Using a virtual 3D Takeshiba, simulation scenarios were inputted along
with data belonging to the area so that the 3D model could be used for
disaster management and other community-development efforts.

2 **Communicating emergency info to improve evacuation behavior**
A function was developed to communicate local conditions via Line. Envisaging an
emergency scenario, the function was used to encourage people in the area to evacuate
effectively during an emergency and to keep them informed of the circumstances,
in order to see whether this would modify people's behavior.

Conceptual images of the pilot

Using PLATEAU

Sta tion

Sta tion

▲ Scenario 1: People head home *en masse*

▲ Scenario 2: People head home in orderly
stages (controlling speed of evacuation
from some facilities)

Tsunami, storm surge, flood risk

Using PLATEAU

Furukawa

Tokyo Bay

Highest watermark
6.0m
3.0m
1.0m

Info broadcasted locally in real time (via Line)

Real time

Integrated data

Information about the vicinity

Compare

General info broadcasted
(as in past emergencies)

Broadcasted over wide geographic area

Broadcasted to individual

Rating: Percentage of people in emergency scenario in whom positive evacuation behavior was induced

Fig. 2.5 Smart-city Takeshiba (Ministry of Land, Infrastructure, Transport and Tourism 2021)

integrating waterborne and overland transport services. Takeshiba's initiatives
reflect the features of the area. For example, with Takeshiba exposed to a threat of a
tsunami or storm surge, a 3D city model was used to simulate congestion conditions
during an emergency (a scenario in which workers in the area attempt to head home
en masse and a scenario in which they head home in stages). The data was also
linked with the messaging app Line to see whether real-time communication of
emergency information could modify people's behavior.

2.5.4 Kibi Kōgen City (Kibichūō, Kaga District, Okayama Prefecture)

Japan's smart-city initiatives are not confined to the country's main cities. They can
also benefit Japan's peripheral (or provincial, outlying, etc.) municipalities, which
are beset by a shrinking and aging population. Technology-driven solutions to these
challenges can, when their effectiveness is demonstrated, create formidable ripple
effects in the peripheral municipalities. The fourth case we explore exemplifies
smart-city initiatives in a typical peripheral municipality. The case is set in Kibichūō,
a municipality in Okayama Prefecture, and the initiatives focus on wellness and
healthcare solutions. Situated in a mountainous region of Okayama Prefecture with
a shrinking and aging population, Kibichūō faces the challenge of delivering community healthcare and advanced emergency services. Designated as a "digital-
garden-health special zone" in 2022, Kibichūō launched smart-city initiatives
coupled with deregulation in an effort to address these challenges. The smart-city

initiatives are designed to "create a futuristic city where residents live in hope, peace of mind, and safety." To that end, they use the municipality's data-linkage platforms to deliver remote healthcare services. Examples are discussed below.

Like many rural communities, Kibichūō lacks an advanced emergency hospital, meaning that people have to travel to another area for emergency treatment or for nighttime pediatric services. Kibichūō aims to address this problem by digitizing ambulances so that emergency procedures can be performed inside the vehicle. Digitized ambulances record video footage of the patient using cameras mounted on the vehicle and camera glasses worn by the crew and then transmit the footage, along with other biodata, to Okayama University Hospital (the region's core hospital). Viewing this footage, doctors at the hospital direct the crew, enabling the swift delivery of emergency care. Kibichūō has also eased the restrictions on paramedics using ultrasound so that more and better medical information can be collected during transit ultimately enhancing the effects of emergency treatment.

Kibichūō is still piloting this scheme to allow paramedics to deliver a greater range of emergency medical treatment under the direction of a doctor, but the Ministry of Health, Labour and Welfare has already started looking into rolling out the initiative. The initiative also involves a paramedic identifying the patient using the patient's ID card (My Number card) and accessing patient data on an emergency data repository (Kibi Concierge Service), which the patient would have previously consented to share. The paramedic would then share this information with the destination hospital as soon as possible. In this way, the initiative involves a plan for using data-linkage platform, and it is on the way toward a full-scale rollout (Fig. 2.6).

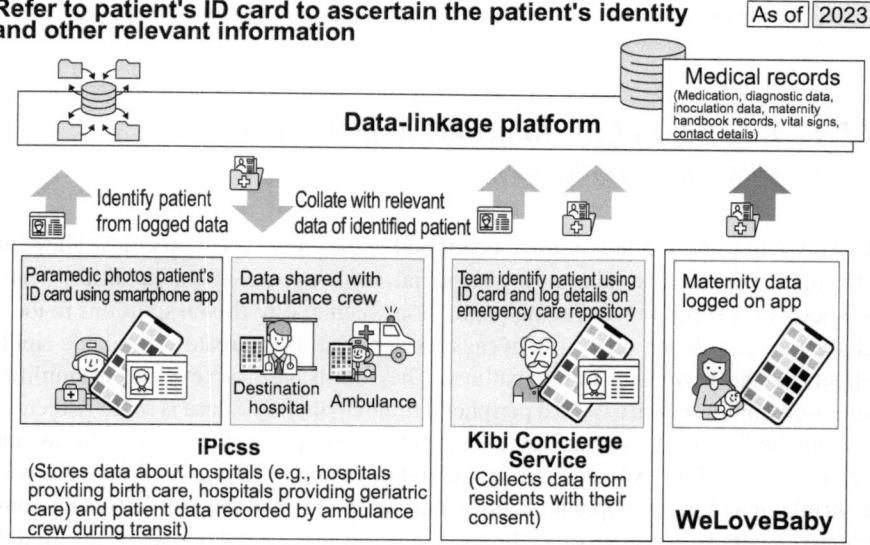

Fig. 2.6 Kibichūō's emergency care services (Office for promotion of regional revitalization, Cabinet Office, Digital Agency 2023)

For residents unable to use private transport, Kibichūō is looking into ride-sharing schemes for hospital visits and for home deliveries of medicine and other goods. This scheme uses digital technology to enhance the welfare of residents: An AI program is used to match available private vehicles with demand; a fleet manager of an existing taxi company checks drivers' ID cards to monitor their health status; and vehicles are equipped with a road-safety rating system to enable ongoing monitoring of the driver's road-safety level.

2.5.5 Taiji Town (Wakayama Prefecture)

Wakayama Prefecture extends down to Nanki—the southern part of the Kii peninsula. The southern part of the Kii peninsula is remote. By air, the region's Nanki–Shirahama Airport is just an hour away from Tokyo. By land, the situation is considerably different; even with decent rail services and expressways, a substantial amount of time is still required to access the region from the big cities of Osaka or Nagoya. In addition to being remote, the region's coastal areas would suffer damage in a megathrust earthquake on the Nankai Trough. Several municipalities in this region have teamed up with an IT vendor to develop down-to-earth solutions for problems related to the aging population.

One of the municipalities in the region is the whaling town of Taiji. Taiji has a population of just less than 3000, and ~ 45% of the population is aged 65 or older. The town has unveiled a vision for a community where elderly people experience fresh joy as they grow older. In a bid to encourage elderly residents to venture outside more, Taiji has launched a community bus service and driverless micro vehicles.

The driverless micro vehicles are fitted with a global positioning system (GPS) tracker that displays their location on a map portal called Elcompass. The GPS was developed by an IT vendor with ties to the community. The map portal was designed mainly with elderly users in mind. As such, the user interface was designed to be accessible to, and to maximize the user experience of, users unfamiliar with digital tools with limited digital literacy. For example, the interface uses a text size, color scheme, and layout to enhance visibility and keeps the information granularity low so that the user can intuitively tell where a vehicle is and the direction it is going. In addition, the portal can be accessed on smartphones and displayed on signage in supermarkets, administration buildings, and hospitals.

In addition to displaying the locations of the driverless vehicles, the portal shows information about evacuation centers, hazards, and sightseeing spots, extending its applicability to disaster management and tourism and increasing the overall capacity to bear costs. The idea is that, by providing mobility options that encourage elderly people to go outside more, the system will reduce future medical costs (Fig. 2.7).

Municipality pioneering the tech		Innovative services		
		Services	Linked data, apps	
Target municipality	Taiji (Wakayama Pref.)	Service provided by Uhuru: Operating system used by driverless vehicles	Sensors	GPS data (for community buses and driverless micro vehicles)
Population	2,903 (resident register as of Oct. 31 2022)		External data	Bus stops
				Evacuation spots
Budget / Year ended March 2023	Digital Garden-City Nation Initiative Promotion Grant Type 1 (Cabinet Office)	Linked external services	Digital signage (to be in seven locations)	

Outline of the services

Aiming to be a place where residents experience life amid the nature outdoors, Taiji has unveiled a vision for a community where elderly people experience new joy as they grow older. This vision will be delivered by providing driverless mobility services in areas with a high elderly population. These mobility services will 1) ensure elderly mobility, 2) encourage elderly people to spend more time outdoors, and 3) be safe and convenient to use.

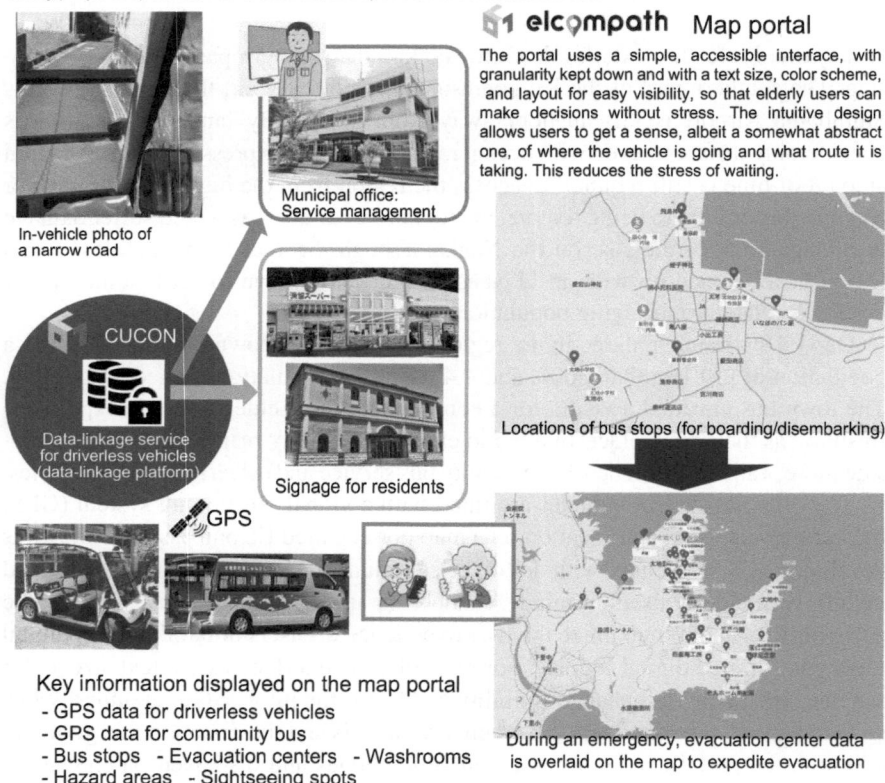

elcompath Map portal

The portal uses a simple, accessible interface, with granularity kept down and with a text size, color scheme, and layout for easy visibility, so that elderly users can make decisions without stress. The intuitive design allows users to get a sense, albeit a somewhat abstract one, of where the vehicle is going and what route it is taking. This reduces the stress of waiting.

In-vehicle photo of a narrow road

Municipal office: Service management

CUCON

Data-linkage service for driverless vehicles (data-linkage platform)

Signage for residents

GPS

Locations of bus stops (for boarding/disembarking)

Key information displayed on the map portal
- GPS data for driverless vehicles
- GPS data for community bus
- Bus stops - Evacuation centers - Washrooms
- Hazard areas - Sightseeing spots

During an emergency, evacuation center data is overlaid on the map to expedite evacuation

Fig. 2.7 Data-linkage platforms rolled out in Taiji (*Translated from visuals provided by Uhuru Corporation)

The aforementioned initiatives illustrate a needs-driven approach to developing smart-city initiatives for addressing the challenges of shrinking and aging populations in outlying regions. They offer plenty of insights to other peripheral municipalities regarding how to identify the problems, what organizational setup to use, and other aspects.

2.5.6   Susami Smart City (Wakayama Prefecture)

Similar to Taiji, Susami is situated in southern Kii and has a population of 3600, with 45–49% aged 65 or older. The municipality has started a smart-city project focusing on disaster management. In 2021, it started piloting a system to digitally manage evacuation shelter provisions and provisions stored in delivery centers, to manage mobile ordering services and delivery operations, and to use precise GPS data to deliver goods by drone.

To maintain cost-effectiveness, the system is used in nonemergency times as a mobile ordering system in which users order goods from roadside stations (*michi-no-eki*). Efforts are underway to expand the system from disaster management to use as a tourism portal.

When I visited Susami myself, I was impressed by the dedication of the mayor, the town hall staff, and the locally rooted IT vendor all working together. Under the mayor's leadership, the project has demonstrated agility and adaptability. Susami stands as a model for developing a smart city in a small municipality without the digital talent of a large city.

2.5.7   Tsukuba Super Science City Initiative (Tsukuba, Ibaraki Prefecture)

Finally, we consider an example that involves a super-city project, that of Tsukuba. A super city is supposed to pioneer a model for a city of the future. Rather than focusing on a particular issue or need, it involves a broad sway of resident-focused initiatives covering many different aspects of city life. Super cities were legislated for in 2020 with the "super-city law" (officially, the Amendment to the National Strategic Special Zones Act). The broad influence of resident-focused initiatives includes those designed to enhance administrative procedures, mobility, logistics, long-term care, education, disaster management, and energy management. At least five categories of these initiatives require data integration (with the use of a data-linkage platform) and, in some cases, deregulation. So far, two cities have been legally designated as super cities. Tsukuba is one. The other is Osaka, which is working to develop flying taxis and other innovations ahead of the 2025 Osaka expo.

Under the Tsukuba Super Science City Initiative, innovations are being developed by a council with representatives from industry, academia, government, and finance. The innovations include improvements to administrative procedures such as the oft-discussed plan for online voting in elections; mobility services, including a community mobility scheme to ensure mobility in peripheral districts and a last-mile transportation scheme for the city center; use of delivery robots and delivery drones; infrastructural and energy management; and infrastructure and services for disaster management, including graphical representations of evacuation centers and disaster damage. The year 2022 saw the piloting of a delivery robot consisting of an

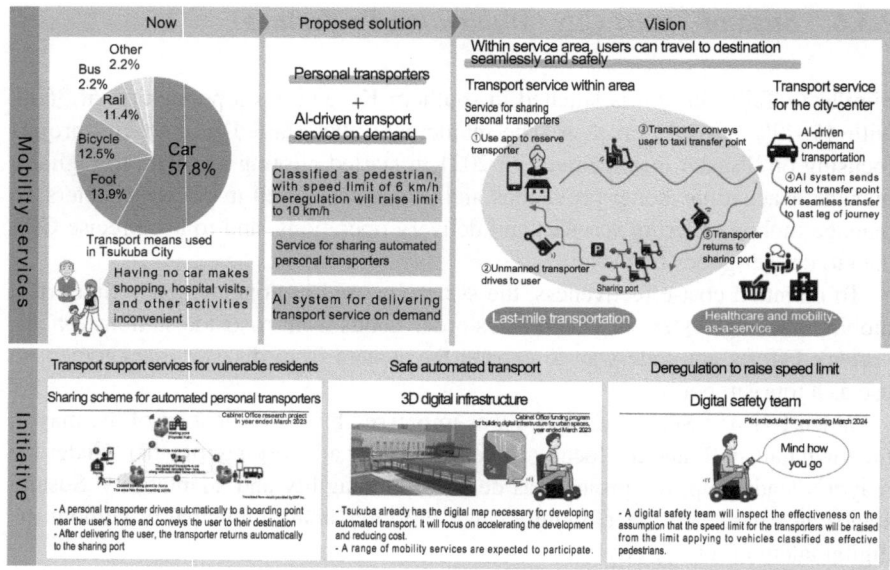

Fig. 2.8 Tsukuba Super Science City Initiative: A new mobility system for delivering mobility and delivery services (Cabinet Office 2023)

unmanned ground vehicle. The robot navigated public roads to deliver goods on demand from supermarkets and cafes, with a delivery fee of 110 yen.

Among the many initiatives, those related to mobility and delivery initiatives that are driven by an innovative mobility system are depicted in Fig. 2.8. With the aging population in regions across Japan, a priority task is to ensure mobility for elderly residents who do not or cannot use a private car. The city has developed a last-mile transportation service in which users book an automated personal transporter.

An app is used to reserve a one-person personal transporter. The transporter drives autonomously to a boarding point near the person's home. It then conveys the user to a taxi transfer point, to which a taxi will be sent by an AI-driven on-demand taxi service. In this way, the user can reach their destination (which might be a hospital, for example) seamlessly, safely, and without stress. Once the user boards the taxi, the personal transporter will navigate its way back to the sharing port, where it will wait for the next callout. The personal transporters Tsukuba City envisages would be classified, under the Road Traffic Act, as pedestrians, meaning that they would have a speed limit of 6 km per hour. However, Tsukuba City hopes to be granted a dispensation that would allow the vehicles to travel up to 10 km an hour.

Tsukuba has also developed some initiatives that reflect the city's unique features. For example, with Tsukuba home to the University of Tsukuba and some national research institutes, the initiatives include a plan to enhance the Tsukuba Startup Park to encourage entrepreneurism among researchers, and a plan to develop grant programs to foster a startup ecosystem.

Envisaging the Desired City for 2050: A Scenario Analysis Using an AI Tool for Policy Recommendations

Local communities across Japan face a host of problems and have launched initiatives to address them. By addressing their present issues, communities will gradually progress toward a better future. It is also important, however, to backcast—to start by envisaging a desired future scenario and then identify what must be done to close the gap between the present scenario and the desired future. Just what is this desired future? H-UTokyo Lab conducted a scenario analysis in an attempt to answer this question. This scenario analysis is outlined here.

The study in question began in April 2020 under the title "Envisaging the Desired City for 2050." After Covid-19 became a pandemic, H-UTokyo Lab had to meet online instead of face-to-face. With this experience of remote meetings, one of the members noted a shift in working practices and moved house to a locale 2 h away from the city center. This locale lay beyond the suburban commuter belt. Many other people in Japan did the same thing, creating a net population outflow from the Greater Tokyo Area to peripheral cities. Media reports highlighted statistical changes in smartphone GPS data and a rise in viewings of homes in peripheral cities (Nikkei 2021). Peripheral cities offer poorer access to big cities than suburban commuter belts do, but it seems that many consider this an acceptable tradeoff given fewer commutes. The longer commute times are apparently more than offset by connection with nature; it seems that people are starting to place more value on this.

This trend may offer fresh insights for a future national geodemographic structure. Until now, two such structures have been considered: a structure in which the population is concentrated into a handful of metropolitan areas, and one in which the population is dispersed among suburban and rural municipalities. To this, a third may be added: expanded agglomerations that encompass rural municipalities, in which urban conglomerations creep outward (Fig. 2.9). Would expanded agglomerations that encompass rural municipalities represent a desirable national geodemographic structure? Which of the three structures— concentration in metropolitan areas, dispersal among suburban and rural municipalities, or expanding agglomerations that encompass rural municipalities —should we aim for? Once we have decided on the desirable scenario, how do we get there? To answer these questions, we ran a simulation and conducted a scenario analysis.

The question of what geodemographic structure is desirable is a nuanced conversation, with multiple perspectives to consider. Considering what is best for employment rates and the economy is important, but considering issues such as carbon footprint and subjective well-being is also crucial. A number of metrics would need to be considered, including unemployment rate, greenhouse-gas emissions (in each case, the lower, the better), and the

(continued)

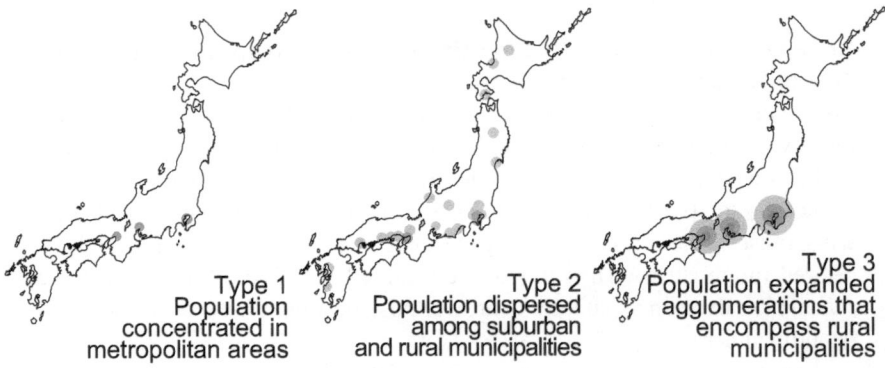

Fig. 2.9 Future national geodemographic structures

percentage of people who are happy with their lives. The next step would be to run a large simulation that would estimate the scores for these metrics in a 2050 scenario. Running an exhaustive scenario analysis was unfeasible; thus, the simulation we conducted focused on the key variables for the metric outcomes and simplified all other variables. In this analysis, we used a simulation tool the Hitachi Kyoto University Laboratory had developed. It has been dubbed the "policy-recommendation AI (Hitachi Kyoto University Laboratory 2020)." Without going into detail here, we inputted a set of metrics aligned with the purpose of the study and ran our own simulation.

Setting the Preconditions for the Simulation: Three Zones

The purpose of the simulation was to compare three hypothetical geodemographic structures for Japan. We grouped Japan's 1700 municipalities into three categories: metropolitan, suburban, and rural. We defined these zones in accordance with the urban employment area definitions provided in Kanemoto and Tokuoka (2002) and used prominently in urban economics in Japan (Kanemoto & Tokuoka 2002).

The metropolitan category encompasses urban municipalities that are located in the urban agglomerations of Tokyo, Osaka, or Nagoya, and in which at least 30% of the working population commutes to the city center. Beyond these municipalities lie suburban communities, defined as municipalities from which a rail commute lasts 90 min from starting station to destination station—our benchmark for a door-to-door commute of up to 2 h. Beyond these lie the rural communities. Under the definitions we used, Yokosuka, Kamakura, Fujisawa, Chigasaki, and Zushi fall under the metropolitan category as they are located in an urban agglomeration of Tokyo, while Hiratsuka, Odawara, Miura, Oiso, Ninomiya, and Hakone fall under the

(continued)

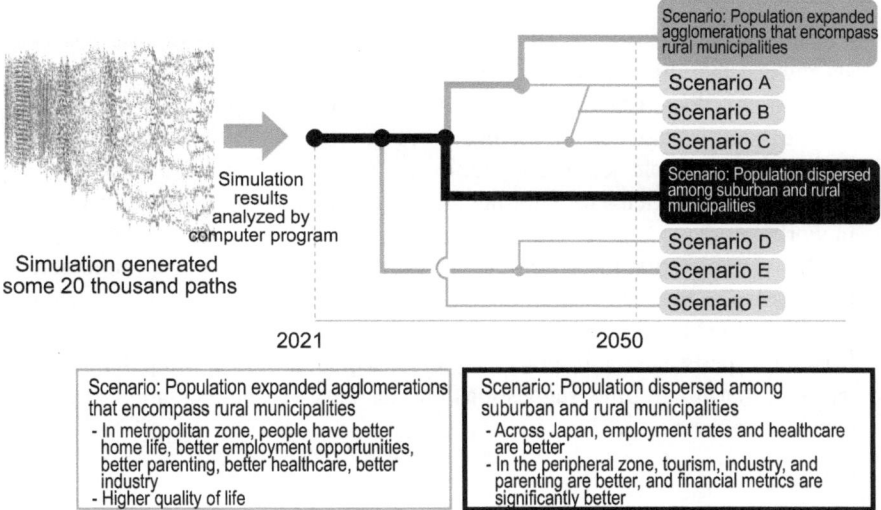

Fig. 2.10   Simulation (by policy-recommendation AI) and analysis

suburban category. We applied the same standard to the urban agglomerations of Osaka and Nagoya.

Having delineated these three broad zones, we then set some 50 zone-specific metrics, including productive population in zone. We also set some 170 national metrics (which apply regardless of zone). With these metrics established, we ran the simulation.

Deciphering the Simulation Outputs

Our AI tool produced tens of thousands of future scenarios. We used a computer program to process these outputs, consolidating them into just a handful of key future scenarios. We then analyzed the metrics for these scenarios to determine which scenario is most desirable, how this scenario can be achieved, and how much time would be required (Fig. 2.10).

The most desirable scenario turned out to be the scenario in which the population is dispersed among suburban and rural municipalities. The expanded suburban-belt scenario was not too bad overall, but it was trumped by dispersal among suburban and rural municipalities. Dispersal among suburban and rural municipalities is associated with better ratings for national metrics such as employment and healthcare, better ratings for rural-specific metrics related to tourism, industry, and parenting, and significantly better ratings for region-specific financial metrics. While we had posited the third scenario (expanded suburban belts) based on our observation of the pandemic-driven exodus to areas beyond the suburbs (as opposed to moving just out to

(continued)

the suburbs, as was the prevailing trend theretofore), the results of our analysis suggest that this scenario is not as desirable as the scenario in which the population is dispersed among suburban and rural municipalities.

The analysis also yielded findings about how to achieve this scenario. One immediately feasible action, for example, is to promote shared transport services and other policies for building a compact city and driving the transition in working practices.

We recognize that these findings are derived from a computer simulation that involved a host of assumptions. As such, the findings should not be taken at face value. Hard conclusions should only be drawn after consultation between experts and the responsible parties. The benefit of a computer simulation is that it can offer suggestions and insights that humans might never consider on their own.

References

Cabinet Office (2023) *Dai 1 kai Tsukuba-shi sūpā shiti gata kokka senryaku tokubetsu kuiki kaigi* [First conference about Tsukuba super city national strategic special zones]. March 16, 2023. https://www.chisou.go.jp/tiiki/kokusentoc/tsukubashi/dai1/shiryou.html. Accessed March 15, 2023.

Council for the realization of the vision for a digital garden city nation, Cabinet Office (2022a) *Dijitaru den'en toshi kokka kōsō kihon hōshi*n [Basic policy for the vision for a digital garden-city nation]. June 7, 2022. https://www.cas.go.jp/jp/seisaku/digital_denen/index.html. Accessed March 15, 2024.

Council for the realization of the vision for a digital garden city nation, Cabinet Office (2022b) Comprehensive Strategy for the Vision for a Digital Garden City Nation (*DIGIDEN*). December 23, 2022. https://www.cas.go.jp/jp/seisaku/digital_denen/pdf/20221223_gaiyou-e. pdf ◦ Accessed March 15, 2024.

Dai-Maru-Yū smart city website. (2024) https://www.tokyo-omy-w.jp/. Accessed March 15, 2024.

Digitalization committee, LDP Policy Research Council (2020) *Dijitaru Nippon 2020: Korona jidai no dijitaru den'en toshi kokka kōsō* [Digital Nippon 2020: Corona Era Digital Garden City Nation]. June 11, 2020. https://jimin.jp-east-2.storage.api.nifcloud.com/pdf/news/policy/200257_1.pdf. Accessed March 15, 2024.

European Commission (2010) Europe 2020. https://ec.europa.eu/eu2020/pdf/COMPLET EN BARROSO%20%20 007 - Europe 2020 - EN version.Pdf. Accessed March 15, 2024.

European Commission (2014) Horizon 2020. https://ec.europa.eu/programmes/horizon2020/en/. Accessed March 15, 2024.

FutureCity Initiative (2019). https://future-city.go.jp/en/. Accessed March 15, 2024.

G20 Global Smart Cities Alliance for Technology Governance (2024). https://www.globalsmartcitiesalliance.org/home. Accessed March 15, 2024.

Hitachi, Sharp, Mitsui Fudosan and Nikken Sekkei (2014) *Kashiwa-no-ha Sumāto Shiti no chūkaku to naru enerugī kanri shisutemu Kashiwa-no-ha AEMS to Kashiwa-no-ha HEMS o kaihatsu, 2014 nen 5 gatsu yori dankaiteki ni unyōkaishi* [Kashiwa-no-ha Area Energy Mgmt System and Home Energy Mgmt System developed as core for Kashiwa-no-ha Smart City; phased launch to commence in May 2014]. April 26, 2014. https://www.mitsuifudosan.co.jp/corporate/news/2014/0424_03/. Accessed March 15, 2024.

Hitachi Kyoto University Laboratory (2020) *Seisaku teigen AI ga egaku Nihon no mirai* [Policy-recommendation AI's depiction of Japan's future]. In: Hitachi Kyoto Univesrity Laboratory (ed) Beyond smart life: *Kōkishin ga kudōsuru shakai* [Beyond smart life: Curiosity-driven society]. Nikkei Business Publications, Tokyo, Chap. 2. Sec. 9.

Info Barcelona, Barcelona City Council (2020) The city's air-quality surveillance and control network is given a boost. June 25, 2020. https://www.barcelona.cat/infobarcelona/en/cerca/the-citys-air-quality-surveillance-and-controlnetwork-is-given-a-boost_964754.html. Accessed March 15, 2024.

Ishida H & Kashiwagi T (ed) (2019) *Sumāto shiti: Society 5.0 no shakai jissō* [Smart city: Delivering Society 5.0]. Jihyosha, Tokyo, pp. 112–121, 182–189

Kanemoto Y & Tokuoka K (2002) *Nihon no toshi ken settei kijun* [Criteria for setting urban zones in Japan] Journal of Applied Regional Science vol. 7: pp.1–15

Kashiwa local Govt website (2024). https://www.city.kashiwa.lg.jp/keiei/shiseijoho/keikaku/machizukuri/kashiwanoha/index.html. Accessed March 15, 2024.

Kashiwa smart city website (2024). https://www.kashiwanoha-smartcity.com/. Accessed March 15, 2024.

METI, Agency for Natural Resources and Energy (2010) *Jisedai enerugī shakai shisutemu jisshō* [Next generation energy and social systems testbed]. https://www.enecho.meti.go.jp/category/saving_and_new/advanced_systems/smart_community/community.html. Accessed March 15, 2024.

Ministry of Land, Infrastructure, Transport and Tourism (2021) Action plan for Smart City Takeshiba. https://www.mlit.go.jp/toshi/tosiko/content/001579884.pdf. Accessed March 15, 2024.

Nikkei (2021) *Chāto wa kataru: Tōkyō kōgai e ijū jiwari: Toshin 100 kiro kennai ni kanshin* [Chart speaks: Creeping migration to Tokyo suburbs: Interest in 100-km radius from city center] January 2021.

Office for promotion of regional revitalization, Cabinet Office, Digital Agency, Cabinet Secretariat, The Cabinet Secretariat's Office for the Council for the Realization of the Vision for a Digital Garden City Nation (2022) *Reiwa 3 nendo hosei yosan dijitaru den'en toshi kokka kōsō suishin kōfukin dijitaru jissō taipu gaiyō* [Overview of digital tech delivery category in digital garden-city nation initiative promotion grant in the revised budget for year ending March 2022]. January 14, 2022. https://www.chisou.go.jp/sousei/about/mirai/pdf/denenkouhukin_jissou_gaiyou.pdf. Accessed March 15, 2024.

Office for promotion of regional revitalization, Cabinet Office, Digital Agency (2023) *Dijitaru den'en toshi kokka kōsō kōfukin (dijitaru jissō taipu) no kōfu taishō jigyō no kettei ni tsuite* [On criteria for deciding eligibility for digital garden-city nation initiative promotion grant (digital delivery category)]. March 10, 2023. https://www.chisou.go.jp/sousei/about/mirai/pdf/dejidenkoufukin_saitaku.pdf. Accessed March 15, 2024.

Secretariat for Promotion of Regional Revitalization, Cabinet Office (2024a) *Sūpā shiti dijitaru den'en kenkō tokku ni tsuite* [About super cities and digital garden health special zones]. January 2024. https://www.chisou.go.jp/tiiki/kokusentoc/supercity/supercity.pdf. Accessed March 15, 2024.

Secretariat for Promotion of Regional Revitalization, Cabinet Office (2024b) Super City Initiative. https://www.chisou.go.jp/tiiki/kokusentoc/english/super-city/index.html. Accessed March 15, 2024.

Sidewalk Toronto (2024). https://www.sidewalklabs.com/toronto. Accessed March 15, 2024.

Smart Cities Mission (2021). https://smartcities.gov.in/about-scm. Accessed March 15, 2024.

Smart City Institute Japan (2024). https://www.sci-japan.or.jp/. Accessed March 15, 2024.

Smart City Public-Private Partnership Platform (2024) https://www-mlit-go-jp.translate.goog/scpf/index.html?_x_tr_sl=ja&_x_tr_tl=en&_x_tr_hl=ja. Accessed March 15, 2022.

UDCK Town Management website. (2024) https://www.udcktm.or.jp/. Accessed March 15, 2024.

UDCK website. (2024) https://www.udck.jp/. Accessed March 15, 2024.

Part II
Approaches to Applying Society 5.0
Architecture in Smart Cities

Chapter 3
Derivation of Key Factors as Methods and Implementation Procedures for Society 5.0 Architecture

Atsushi Deguchi

Abstract This chapter explores how the Society 5.0 reference architecture can be applied to the development of smart cities, emphasizing a people-centric, integrated approach. Society 5.0 envisions a new model for urban design where advanced technologies and data-driven strategies converge to create cities that enhance quality of life, economic opportunities, and environmental sustainability. Building on the foundational elements of modularity, interconnectivity, and adaptability, this framework supports the creation of smart-city infrastructures that are resilient, efficient, and responsive to citizens' needs. To fully realize this vision, three essential components are identified to complement the reference architecture: streamlined processes for efficient service delivery, interfaces that prioritize ease of use and accessibility, and organizational frameworks that foster collaboration across public and private sectors. In this context, these components serve as the foundational elements, which structure the methods and means for implementation. Subsequent chapters will discuss six specific key factors, further positioning them within the smart-city framework. By integrating these elements, Society 5.0's architecture provides a comprehensive blueprint for transforming urban spaces to harmonize technological advancements with citizens' daily lives, ultimately leading to sustainable and vibrant communities. This integrated approach not only improves the quality of urban living but also fosters innovation, resilience, and inclusivity, paving the way for future-ready cities.

Keywords Reference architecture · Implementation roadmap · Process · Interface · Organization

A. Deguchi (✉)
Department of Socio-Cultural Environmental Studies, Graduate School of Frontier Sciences, The University of Tokyo, Tokyo, Japan
e-mail: deguchi@edu.k.u-tokyo.ac.jp

© The Author(s) 2025
Hitachi-UTokyo Laboratory (H-UTokyo Lab.), *The Architecture of "Society 5.0"*,
https://doi.org/10.1007/978-981-96-2929-9_3

3.1 Intentional Interpretation of the Society 5.0 Reference Architecture

Recall the topic discussed in the first chapter: To create smart cities (which relate to community development) and smart factories (which relate to manufacturing) that align with the Society 5.0 vision, we need an architecture that describes the overall design. In that chapter, I presented the broad generic architecture based on the government's Society 5.0 reference architecture (Fig. 3.1) (Cabinet Office 2020) and discussed the way of thinking and approach that the architecture represents. I discussed the significance of the Society 5.0 reference architecture, with its eight tiers ranging from strategy and policy to assets. I then argued that if the architecture is supposed to be a design, then a construction method and implementation roadmap must be complemented to achieve the design.

In Chap. 2, I outlined the smart-city trends in Japan, illustrating how smart cities are one way of embodying the Society 5.0 vision. If smart cities are mapped onto the Society 5.0 reference architecture, they would represent a single layer on the "domains" axis (the depth axis extending into the background).

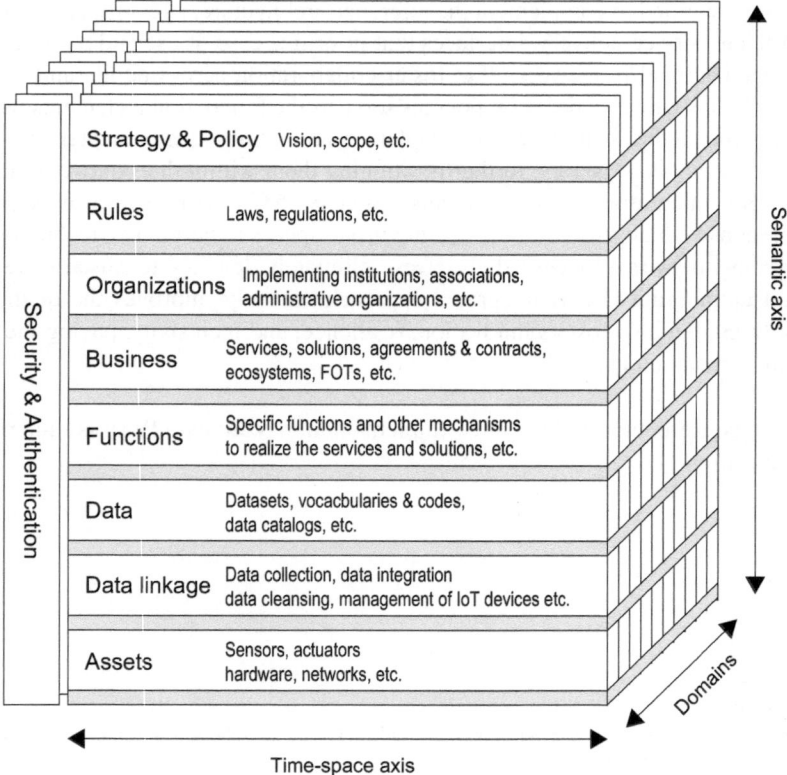

Fig. 3.1 Society 5.0 reference architecture

We shall assume that each layer in the "domains" stack represents a domain that embodies the Society 5.0 ethos. We then assign each layer a label that roughly represents that domain. One of the layers shall be labeled "community-development (smart city)" (admittedly, the relationship between smart cities and community development is, in many ways, extremely complex for us to consider a smart city synonymous with smart [digitized] community development). Other layers can include "manufacturing (smart factory)," "healthcare (smart healthcare)," "finance (smart finance or fintech)," "energy (smart energy)," and "mobility (smart mobility)" (Fig. 3.2).

Services and business were previously arranged vertically and horizontally; however, with this eight-tier reference architecture, we can re-envisage the services and businesses as being arranged in the Society 5.0-embodying domains, which, in the figure, run along the "domains" depth axis. The architecture for a Society 5.0-embodying smart city, for example, becomes one of the domain layers (labeled "smart city"), which, similar to all the other domain layers, has an eight-tier reference architecture.

However, another look at the reference architecture in Fig. 3.1 reveals that something may be missing. If Society 5.0 is supposed to be people-centric, where is the user? The user's absence is perhaps unsurprising. The Society 5.0 reference architecture is presented from the perspective of the service providers and the manufacturers. As the architecture presents a broad general structure that can allow data security and data interoperability between different industries and operators, the

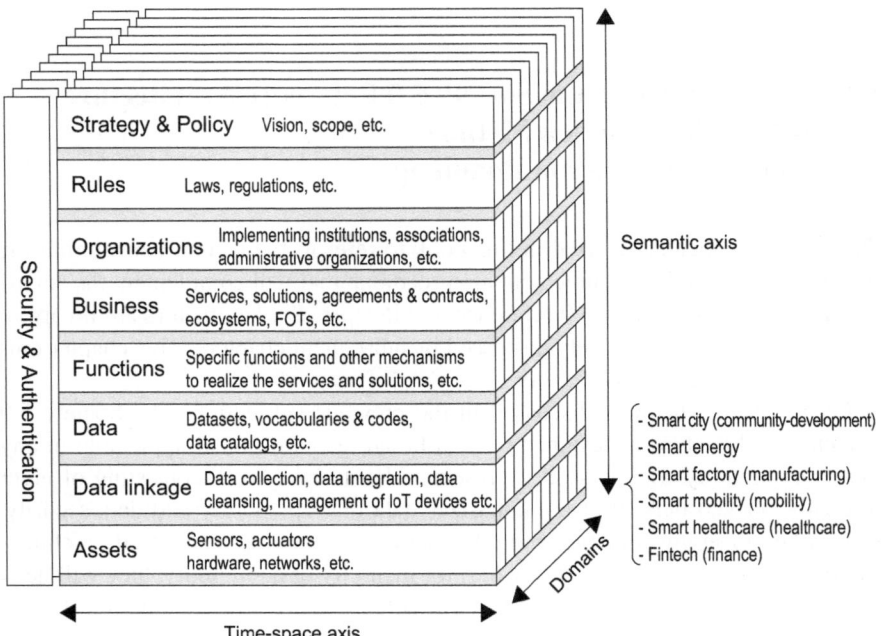

Fig. 3.2 Society 5.0 reference architecture with domain labels added

architecture omits cities, communities, and residents—the people—who use the services or receive the goods. As the eight tiers omit the role of cities, communities, and people, and given that we need an architecture that can deliver people-centered smart cities in real-world cities and communities, then something must be done to link the reference architecture with the lives of people in these real-world cities and communities.

Recall the discussion in the second half of Chap. 1 about the need for a construction method and implementation roadmap. After outlining the eight-tier structure of the reference architecture, I argued that when a smart-city initiative in an actual city or community needs to be delivered based on blueprints for each smart-city initiative, a construction method and implementation roadmap must be set out. I emphasized that the key to smart cities lies in having a construction process and implementation roadmap aligned with the people-centrism principle of Society 5.0. To reiterate, each smart-city initiative will have its blueprints setting out the vision to be achieved, but to deliver the blueprints in reality, there needs to be a construction method and implementation roadmap that aligns with the city or community's features and with the residents' values. This is a common requisite for any smart-city initiative.

If the necessity of a construction method and implementation roadmap is a common factor across all smart-city initiatives, what are the key factors behind this, and how do we complement the architecture? This is the very question that drove the research upon which this book is based (research on the key factors presented in Chap. 4 onward).

3.2   Complementing the Society 5.0 Reference Architecture with a Construction Method and Implementation Roadmap

The term "key factors" is used in this book to refer to the factors constituting the construction method and implementation roadmap that will complement the smart-city domain in the Society 5.0 reference architecture. Six key factors are involved, and these are comprehensively discussed in subsequent chapters. This chapter outlines the background to the factors.

Let us recap what was discussed in the previous subsection. The Society 5.0 reference architecture is designed to be a broad, general model applicable in any city or community, and as such it presents a basic framework consisting of eight vertically stacked tiers. It also emphasizes data interoperability and data security. Thus, when local governments and other actors are devising a smart-city architecture (one of the domains of Society 5.0) for their city or community, they will need to complement the reference architecture with a construction method and implementation roadmap. What exactly needs to be added to make the architecture more useful to the local government and other local actors (businesses, for instance) responsible for creating a smart city?

To determine the kind of construction method and implementation roadmap that must be added to the smart-city domain, we should first add to the reference architecture the actors—city, community, people. We can then try to identify the necessary factors. Remember, the Society 5.0 architecture omits the people who receive or use the services and who live their lives in the city or community in question. That is why we add the "city, community, people" of the smart city in Fig. 3.3.

Although the reference architecture was originally oriented around the perspectives of parties making the smart city and delivering the services, by adding "city, community, people" (the parties on the receiving end) to the figure, we can incorporate the perspectives of those who receive or demand the services—the users of the smart city. In turn, when the architecture has incorporated the perspectives of the user/demand side and the city and community where the people live their lives, people-centrism and sustainability of the city or community will be considered in the design of the construction method and implementation roadmap. Then, when the architecture is finally complemented with a construction method and an implementation roadmap emphasizing people-centrism and sustainability, devising a more practical architecture for building a people-centric sustainable smart city will be possible.

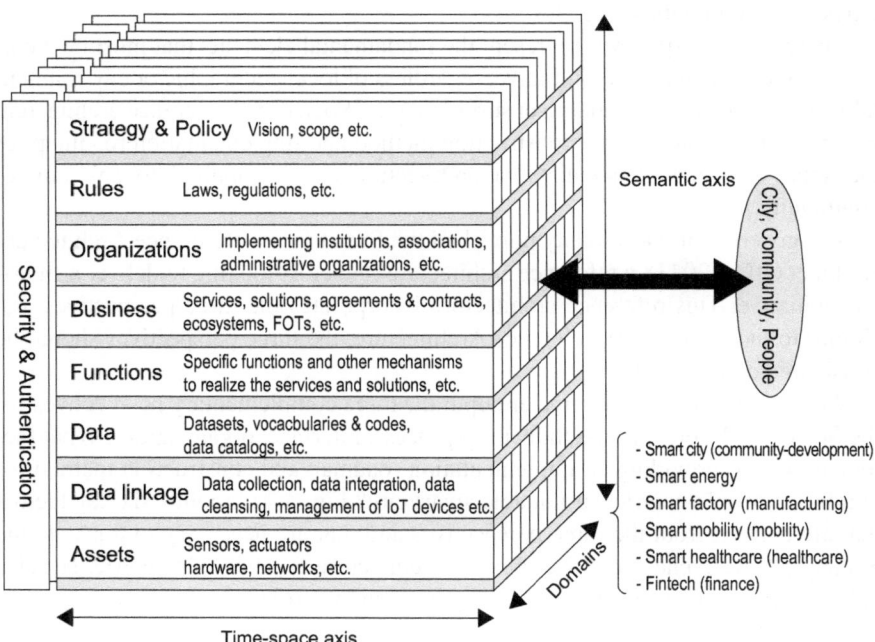

Fig. 3.3 Superimposing "city, community, people" to the Society 5.0 reference architecture to incorporate the perspectives of the user

Once we have added "city, community, people" to the figure, an important perspective should be revealed: across the "assets" layer (the bottommost of the eight tiers), we need to add an interface to represent the data-exchanges, communications, and service exchanges that mediate the delivery of the assets (in the bottommost layer of the eight-tier architecture) to "city, community, people" (Fig. 3.3).

The time–space axis, labeled at the bottom of the figure, represents, albeit in a simple way, an important perspective for the delivery process: the time taken to undertake the series of delivery processes for implementing the smart-city plan and for developing and introducing the services across a larger area. With this perspective in mind, all the greater care will be taken when drafting the implementation roadmap. The delivery process is critical for the wrong delivery process could create irreversible problems.

Another vital perspective for the construction method and implementation roadmap is the organizational setup—developing an organizational infrastructure and allocating experts and businesses in a way that aligns with the community's features.

To recap, we have identified three important elements: an interface that connects the eight layers of the reference architecture with the actual city, community, and people; a process for delivering and scaling up over time and in stages; and an organizational setup detailing the organizational and procedural structures for coordinating a unified community effort. These three perspectives need to be added when the Society 5.0 reference architecture for the smart-city domain is applied in actual cities and communities (Fig. 3.4).

These three perspectives represent the fundamental elements that must be incorporated into any smart-city initiative, regardless of local geographic or environmental factors. They are essential to ensuring that the Society 5.0 reference architecture can be supplemented with a construction method and implementation roadmap for the smart-city domain so that the architecture can be applied to any city or community.

The government has already published the Smart Cities Reference Architecture (Cabinet Office 2023), a reference architecture for actors looking to deliver a smart-city initiative. This reference architecture incorporates the three perspectives, but similar to the Society 5.0 Reference Architecture, the three perspectives should be considered complementary.

We shall now explore in greater depth the three complementary perspectives for the Society 5.0 Reference Architecture (process, interface, organization). What factors are key to providing an implementation roadmap and construction method for building a people-centric, sustainable smart city? Fig. 3.4 pinpoints six key factors that are derived from the three perspectives and that are necessary to translate the Society 5.0 Reference Architecture into architecture for a people-centric, sustainable smart city. It also shows where they fit in and the role they play.

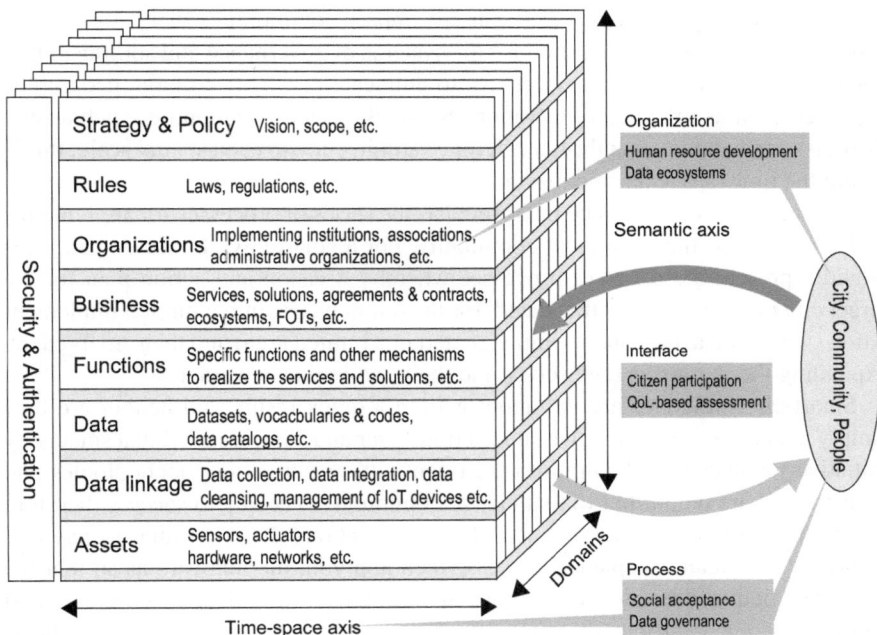

Fig. 3.4 Society 5.0 reference architecture supplemented with the perspectives and key factors for "city, community, people"

3.3   Three Perspectives and Six Key Factors Added to the Architecture

3.3.1   Key Factors for Process

No one can make a Society 5.0-embodying smart city overnight; it takes time. The time factor is represented in the reference architecture by the time–space axis, but to properly understand the time required, we need to break this axis down. For this axis, we focus on the delivery process. Given that a smart city is not a short-term project that would take a year or a few years to complete, the procedures and steps taken in the delivery process will shape the future of the project.

In the case of the smart-city domain, the time–space axis could be interpreted in a number of ways, but here we will approach it from the perspective of urban planning. Just as an urban plan can have multiple subdivisions or units reflecting the spatial scale of the target area, the same applies in the case of a smart city; the planning needs to reflect the spatial scale of the target area, which could be an expansive geographic region (such as a prefecture or large urban agglomeration), a provincial municipality (such as Hamamatsu or Aizu-Wakamatsu), or city district (such as Kashiwa-no-ha Smart City or Dai-Maru-Yū Smart City).

As the "society" in Society 5.0 can encompass communities of various scales, separate architectures are required for the systems of governance and administration particular to each scale. It is necessary to delineate scale—from large geographic expanses such as urban agglomerations to smaller-scale communities such as city districts—and apply the eight-tier reference architecture to each spatial scale; this is where the time–space axis comes in.

The time–space axis can be interpreted as the series of processes for applying the eight-tier architecture to a local community. For a smart city, the time–space axis would represent the series of processes whereby a vision and action plan for the target city or community is translated, for the initial phase, into a limited implementation (where the action plan is implemented in a limited area) and then enlarged, by expanding the initiatives themselves and their geographic coverage.

Smart-city initiatives are not about getting a quick result, and neither are they simply a matter of repeating the same actions over and over. Delivering a smart-city initiative involves introducing into the target city or community technologies and systems that are unfamiliar to that city or community. The key to success therefore lies in the community's acceptance of the innovations. It is essential to pay due regard to the social acceptance process when applying the innovations so that the initiatives tied to the new technologies and systems will be locally grounded and therefore people-centric and sustainable. Because the technologies will be collecting and using personal data along with other community data, robust rules and data governance will be critical to obtaining social acceptance. Therefore, under the process perspective, we placed social acceptance together with data governance. The two factors are discussed in Chaps. 5 and 6.

These two factors must accompany the implementation of the services and businesses related to the smart city.

3.3.2 Key Factors for Interface

I mentioned earlier that once we place "city, community, people" in the Society 5.0 Reference Architecture as shown in **r**, we will also need to add an interface linking reference architecture with the "city, community, and people" together to ensure that the services are delivered in a way that aligns with their values.

The reason is that as the smart city is supposed to provide digitally driven solutions for the city or community's problems, it is necessary to ensure communication between the supply and demand side—between the service providers (the local government and enterprises delivering the businesses or services) and the service users (the citizens, consumers, or residents). The question becomes how to introduce and embed into the local community a structure that interfaces between the supply and demand side and monitors outcomes of the services to see whether they are effective, whether any adverse effects exist, and whether they can be improved.

Thus, the construction method and implementation roadmap to be added to the architecture must include an interface mediating between the actors responsible for the smart-city architecture's business (services) layer, functions layer, data layer, data-linkage layer, and assets layer (these actors will usually be the local government and enterprises) and the "city, community, people" (primarily, the local citizens).

One critical factor for this interface is citizen participation. We already have an example of a system for citizen participation—the living labs of Scandinavia.

In addition to needing citizen participation, an interface requires ongoing monitoring of each individual, to as precise a degree as possible, to understand how the smart-city technology is affecting people's lives, how much it is contributing to satisfaction and quality of life among the target citizens, and whether it is creating adverse effects.

Further measures must be taken, then, to precisely and accurately evaluate outcomes in the target population and encourage feedback from them. Urban planners usually take a broad, quantitative approach, using physical, objective measures that give an overhead snapshot (such as green space or road space as a proportion of total city space). However, this quantitative approach is unsuitable for measuring the effectiveness of digitally powered services. Another approach is to use statistical data to find average rates for the city as a whole. However, this approach ill suits the people-centric ethos, which would require consideration of the diverse array of individual preferences. One cannot just rely on government statistics. For the interface linking the supply and demand side, the evaluative approach needed is one that takes into account human diversity and diversity of preferences. Thus, the second key factor for the interface is a system for evaluating the quality of life—QoL-based assessment.

We have, therefore, identified citizen participation and smart-city QoL-based assessment as two elements that will serve effectively as parts of the smart-city procedures and methods to be added to the Society 5.0 Reference Architecture and as key factors in the interface mediating between the parties delivering the services and the citizens using them. The first of these factors is discussed in Chap. 7 and the latter in Chap. 8.

3.3.3 Key Factors for Organization

One question is common to all domains in the top three layers of the Society 5.0 Reference Architecture (these layers being strategy and policy, rules, organizations): the question of who is responsible for determining the elements of the layer in question. The paramount concern here, needless to say, is the competence of whoever sets the strategy and policy, whoever sets the rules, and whoever designs the organizations. For these layers, people with strategic thinking skills and the expertise to tailor rules to the needs and circumstances on the ground are needed in addition to people with the expertise for the business and data layers further down the architecture.

In conventional urban planning, the strategy and policy layers are the responsibility of urban planners who were schooled in urban planning or are otherwise experienced in such. In the case of smart cities, however, this layer requires expertise in the relevant digital technologies and in data-driven services. As such human resource is unlikely to develop of its own accord, a new training program to nurture it is needed. The first key factor, then, is human resource development, representing the need to nurture the human resource necessary to coordinate smart-city collaborations between academic, private, and government actors. Human resource development is discussed in Chap. 9.

A smart city requires public–private collaboration. It cannot be built by government or by private enterprises alone. It must also involve citizen participation. It is, therefore, necessary to allocate the human resource who can create a collaborative organizational framework suitable for the target community and who can coordinate between the different organizations within this framework, so that the smart city is developed in line with the strategy and the businesses and services are developed in a coordinated way.

Here, it is worth clarifying the meaning of the prefix "super-" ("above, over, and beyond") as used in the "people-centric, super-smart society" that is Society 5.0. One meaning of super-smart is a level of smartness above or more advanced than a normal level of smartness. If a "smart" city involves data and digital applications in one industry or sector, then a "super-smart" city involves data and digital applications that are of a higher level and that span different sectors and link them together. Tellingly, the Society 5.0 Reference Architecture includes under the data layer a data "linkage" layer, illustrating how important inter-sector integration is in the Society 5.0 vision. An important part of Society 5.0 is linking data across different sectors to create a data ecosystem in which data mediates between different industries. To ensure that smart-city initiatives are sustainable, it is essential to nurture and facilitate collaboration between the business operators who will keep the smart city running. The key to this is a data ecosystem.

With data as the fount of businesses (or services), nurturing a data-linked inter-industry ecosystem will enable the services to transcend the smart-city domain and cross into other domains running into the background in the figure. Suppose, for example, that CCTV data used in one smart-city domain were combined with foot-flow data and sales data from roadside shops; in addition to encouraging the use of the data in marketing analytics, this would enable the data to be employed in the smart mobility domain (in services designed to ease congestion, for instance). Suppose that the CCTV data were combined with the data from eco-sensors placed on roads; this could spur the creation of advanced services that help mobility-challenged people venture outdoors. Thus, complementing the organization perspective in this way would deepen the connections in the upper layers within the smart-city domain itself and also deepen the inter-domain connections along the "domains" depth plane of the Society 5.0 Reference Architecture.

Thus, data ecosystem is a key factor of the organization perspective. It is discussed in Chap. 10.

3.4 Addressing the Challenges Associated with a People-Centric Sustainable Smart City

Smart cities are bespoke creations, individually tailored to the city or community concerned; they are not to be mass produced with a one-size-fits-all design. Each city or community has its own set of features and environmental conditions, along with its own array of needs and values. The geographic, social, and environmental features of Central Tokyo differ from those in municipalities situated in the peripheries of the Tokyo agglomeration, which themselves differ from those in rural municipalities (farming villages and fishing villages, for instance). Such locality-specific contexts cannot be ignored, and neither can a smart city be built on vacant land.

Every smart city cannot be developed with a single unified process, as goods were during the industrial age (Society 3.0), which emphasized manufacturing and mass production; even if we could, we might end up depriving cities and communities of their unique features, as the modernist and functionalist trends of the twentieth century were criticized for doing.

Each community also has its own set of needs and problems. True, multiple communities sometimes share the same kinds of problems, which might mean that the same initiative can be rolled out horizontally. However, an approach that works in one community can, at best, work in just a few other communities. In addition to targeting problems or needs particular to the community in question, devising an architecture suitable for the community and its environment is necessary, along with a construction method and implementation roadmap for this architecture using the Society 5.0 Reference Architecture as a guide. Chapters 4, 5, 6, 7, 8, 9, and 10 present the key factors for devising a construction method and implementation roadmap for architecture localized to a particular community and environment.

Remember, we must never let smart cities end up a passing fad. The actors seeking to build a Society 5.0-embodying smart city (one that is people-centric and sustainable) must commit to building it in stages over time, and this will require an overarching architecture.

The six factors discussed in the ensuing chapters are critical for creating a construction method and implementation roadmap that will complement the Society 5.0 Reference Architecture, making it applicable to any city or community. As such, they are essential to the phased construction of a smart city that embodies Society 5.0 by being people-centric and sustainable.

The key factors must be understood to ensure that smart-city initiatives can be sustainable and people-centric by contributing to the well-being of residents and do not become a passing fad. They offer local governments and other actors insights about the problems they will need to overcome when devising the architecture for a Society 5.0-embodying smart city and delivering services related to their smart-city project.

In summary, the discussion in this chapter was based on the premise that a local government, enterprise, or other actor involved in building a Society 5.0-embodying smart city should devise the smart city in the order shown below; what is shown below summarizes the discussion in this chapter.

1) Clarifying and communicating the vision
Clarify what the vision is and communicate it to the relevant parties
To that end, refer to the government's Science and Technology Basic Plan and H-UTokyo Lab's *Society 5.0: A People-centric Super-smart Society* (H-UTokyo Lab 2020).

⇩

2) Formulationg an Architecture
Formulate an architecture for the "smart city" domain in the Society 5.0 Reference Architecture
To that end, refer to the Society 5.0 Reference Architecture (Cabinet Office 2020) and Smart Cities Reference Architecture (Cabinet Office 2023)

⇩

3) Developing a construction method and implementation roadmap
Complement the Society 5.0 Reference Architecture with a construction method and implementation roadmap
The challenge here is determining how to make the smart city people-centric and sustainable
To that end, refer to Chaps. 4, 5, 6, 7, 8, 9, and 10 of this book (which discuss the six factors)

If you are involved in managing the delivery of a smart-city initiative or designing the smart-city architecture, perhaps the current progress in your smart-city initiative should be reviewed from the perspective of Society 5.0, what needs to be done should be determined, or what inadequacies need to be rectified should be identified, to bring the project closer to the ideal of a people-centric, sustainable smart city.

This chapter presented a discussion on why the six key factors are necessary for making a people-centric sustainable smart city. Chapters 4, 5, 6, 7, 8, 9, and 10 describe the six factors in depth. We hope that the value and effects of the six factors are understood and then their descriptions, namely, the thinking behind them and their technological aspects, can be applied in your smart city.

The six factors are soft infrastructure for a digital society. Each constitutes a piece of infrastructure for a people-centric, sustainable smart city, and the infrastructure should be developed in stages.

References

Cabinet Office (2020) Smart City Reference Architecture White Paper, 1st edn. In: Cross-ministerial Strategic Innovation Promotion Program (SIP) Second Phase, Big-data and AI-enabled Cyberspace Technologies /Smart City Architecture Development /Smart City Architecture Design and Promotion of Related Verification Research (Released on March 31, 2020). https://www8.cao.go.jp/cstp/stmain/20200318siparchitecture.html. Accessed on March 7, 2024.

Cabinet Office (2023) *Sumāto Shiti: Rifarensu ākitekucha howaito pēpā* [Smart city: Reference architecture white paper], 2nd edn. In: Cross-ministerial Strategic Innovation Promotion Program (SIP) Second Phase, Big-data and AI-enabled Cyberspace Technologies /Smart

City Architecture Development /Smart City Architecture Design and Promotion of Related Verification Research (Released on August 10, 2023). https://www8.cao.go.jp/cstp/stmain/20230810smartcity.html. Accessed March 7, 2024.

H-UTokyo Lab (2020) Society 5.0: A People-centric Super-smart Society. Springer, Singapore. https://www.springer.com/gp/book/9789811529887. Accessed on March 7, 2024.

Chapter 4
Six Key Factors for Making a Smart City People-Centric and Sustainable

Tomoyo Sasao and Shin Osaki

Abstract This chapter examines six key factors essential for the development of people-centric and sustainable smart cities, in alignment with the principles of Society 5.0; these factors include social acceptance, data governance, citizen participation, quality-of-life (QoL)-based assessment, human resource development, and data ecosystems. Each factor is explored in the context of the common challenges encountered in smart-city implementation. Social acceptance addresses the need to build community trust in new technologies, whereas data governance focuses on safeguarding personal data, which is vital for public confidence. The role of citizen participation is emphasized, aiming to promote active community engagement in the planning and implementation of smart-city initiatives. The QoL-based assessment shifts the evaluation criteria for smart-city outcomes from service performance metrics to a focus on enhancing residents' well-being. Human resource development outlines the need for interdisciplinary teams to manage smart-city projects, and the data ecosystem highlights the importance of integrated cross-sectoral data use for sustainable urban environments. This chapter aims to provide a comprehensive understanding of how these factors can be operationalized in real-world communities, ensuring that smart cities align with the people-centric ethos of Society 5.0. The subsequent chapters delve into detailed research findings and offer insights into the practical application of these factors.

Keywords People-centric smart cities · Sustainable urban development · Society 5.0 · Data governance and privacy · Citizen participation and co-creation

T. Sasao (✉)
Faculty of Engineering, Reitaku University, Chiba, Japan
e-mail: tsasao@reitaku-u.ac.jp

S. Osaki
Neighverse Inc., Tokyo, Japan

Sustainable Society Design Center, Graduate School of Frontier Sciences,
The University of Tokyo, Tokyo, Japan
e-mail: osaki@edu.k.u-tokyo.ac.jp

© The Author(s) 2025   55
Hitachi-UTokyo Laboratory (H-UTokyo Lab.), *The Architecture of "Society 5.0"*,
https://doi.org/10.1007/978-981-96-2929-9_4

4.1 For Smart Cities that Manifest Society 5.0

Chapter 2 discussed the background to smart cities, including how the 2000–2010 period emphasized energy-efficiency solutions, solutions for residents' problems, and efficient urban development, how the focus shifted to resilience and survival during the pandemic, and how the focus has now shifted to how smart cities can embody the Society 5.0 vision's people-centric ethos by contributing to well-being, involving citizen-led initiatives, and involving mutual assistance. However, how many smart cities actually are aligned with the Society 5.0 vision?

Chapter 3 discussed the need to map "city, community, people" onto the Society 5.0 Reference Architecture and presented six key factors for the construction method and roadmap that complement the reference architecture. What happens when the six key factors are paired with the problems that arise in smart-city settings that embody "city, community, people" (Fig. 4.1)?

Each of the key factors corresponds to a different problem. For example, social acceptance is key to addressing public resistance to a sudden rollout of a technological innovation, whereas citizen participation is key to addressing the lack of opportunities for citizens to participate in co-creation (because government and private enterprises are taking the lead in managing the smart-city initiatives). The problems corresponding to the six key factors are common to any city or community where a smart city is being created.

In actual smart-city settings, which key factor corresponds to which common problem, and how does one go about addressing the problem? This chapter outlines the common problems to which the six factors pertain and the approach to addressing the problems. Chapters 5, 6, 7, 8, 9, and 10 go into greater depth.

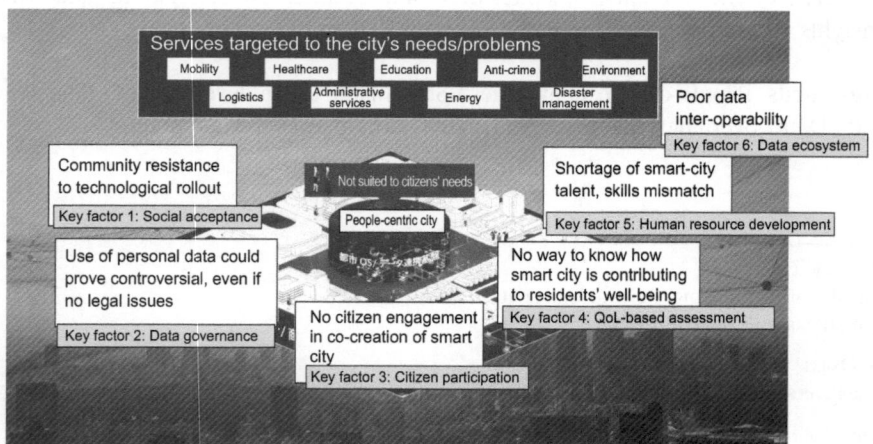

Fig. 4.1 Six key factors for Society 5.0 and their corresponding problems

4.2 Key Factor 1: Social Acceptance

When a smart-city initiative is launched, the first issue encountered is whether the community will accept the technologies and services being introduced. H-UTokyo Lab has therefore identified social acceptance as a key factor and researched the nature of the matter.

Recent smart-city initiatives have raised the prospect of rolling out individualized services driven by granular personal data, including information about people's health and behavior. The problem here is that citizens might object to such personal data being collected. It is also an ethical minefield; for example, collecting and using someone's data without their permission is morally questionable. Around the world, we can find examples of litigation related to this.

We use the term "social acceptance" to refer to a situation in which the community is well-informed about the smart-city initiatives, actively participates in the decision-making process, and welcomes the initiatives. Social acceptance is absolutely crucial to smart cities. Ensuring social acceptance is not just about meeting legal or regulatory criteria for personal data use; it requires a tailored approach to building trust among community stakeholders. Therefore, the approach must be tailored to the needs of the community in question. On the other hand, some features are common to all communities and some ethical matters should be considered in all cases. We have researched these common features with a view to raising awareness knowledge about them. Chap. 5 presents the research findings.

4.3 Key Factor 2: Data Governance

When using data for a smart-city initiative, the first hurdle to overcome is how to handle the data. Using personal data, for which privacy is paramount, could prove controversial even if legal requirements are met. Only when the public feels confident that the organization will safeguard the data will it also have confidence in the organization's ability to run the smart-city initiative. That is why we have identified data governance as a key factor and researched the risks and other matters associated with using people's data.

Smart-city initiatives often involve data- and AI-driven solutions for local problems. The initiatives rely on residents' personal data along with nonpersonal data such as data related to mobility, energy, and other urban services. The data must be properly governed.

In the case of personal data, data governance means, first of all, compliance with legal requirements such as those set out in Japan's Act on the Protection of Personal Information. Second, it means something more than meeting the legal requirements. If the target audience feels that the service violates privacy (though it meets all legal requirements related to data protection), then they would feel reluctant to provide their data, and the service—being predicated on the provision of such data—would be jeopardized. Simply complying with the law is not sufficient; smart cities' designs must incorporate robust data governance to prevent controversy and loss of confidence. The key factor is discussed in depth in Chap. 6.

4.4   Key Factor 3: Citizen Participation

Residents and workers should be co-creators of smart cities. Rather than just passively using technology or receiving services, they should have a say in the future of their own community, actively engage in the decision-making process, and in some cases participate in the delivery process. That much is clear. Less obvious, however, is exactly how to create opportunities for such citizen participation in the smart city. That is why we have identified citizen participation as a key factor and used a living-lab approach to explore ways of facilitating co-creation with the public.

It is residents, workers, and other members of the public who know the community's problems the best. Smart-city initiatives should therefore provide platforms where citizens can share their knowledge, experiences and creativity, fostering co-creation with other stakeholders. The reality, though, is that only very few intersecting spaces are available for the public to engage with smart cities, indicating that in many cases, the public has a poor grasp of the smart-city initiative or feels it is irrelevant to their lives.

To rectify this, the actors in charge of delivering the smart city should provide opportunities, systems, and coordinating functions so that local residents and local organizations will actively engage in co-creating the smart city; they should also ensure an ongoing process of dialog and feedback with the public. This is absolutely vital. One approach to this task is to provide a living lab—an open innovation process in which the public gets involved in the research and piloting. A living lab is similar to a smart city in that it facilitates open innovation in research, in testing, and in piloting innovations in living settings, meaning that the living lab is all the more valuable when tied with smart-city initiatives. Chapter 7 presents our research findings on this key factor.

4.5   Key Factor 4: QoL-Based Assessment

Once the smart-city initiatives start to be delivered, people will start experiencing the new services in their daily lives. The party in charge of the delivery will then start assessing outcomes. To date, the delivering parties have tended to track the progress of their projects or measured the degree to which their set objectives have been attained—they want to know details such as whether the innovation has been successfully delivered, how many users have subscribed to the service, and so on. What they should really be measuring is how the delivery is contributing to people's quality of life or well-being. That is why we have identified QoL-based assessment as a key factor and developed a new assessment metric, ones that closely tracks changes in quality of life that result from the application of personal data. Quality of life and public satisfaction are already assessed in questionnaires conducted every few years by national and local governments. However, this conventional approach is ill-suited to measuring the outcomes of smart-city initiatives, which involve fast-paced changes and developments.

With the dissemination of smartphones, wearables, and IoT-driven tech, it has become easier than ever to collect continuous data from people in real time. Thus, when measuring the outcomes of applying such technologies in a smart-city project, it will be increasingly important to have an assessment model consisting of people-centric metrics such as the quality of life and to have an environment that enables a people-centric approach to assessing individual measures or general outcomes of smart-city initiatives. Chapter 8 presents our research findings on this key factor.

4.6   Key Factor 5: Human Resource Development

Once smart-city initiatives start being scaled up, bolstering the team managing the smart city as a whole will be necessary. The question, though, is what competences, skillsets, and knowledge are needed to guide the overarching smart-city strategy. Another question is what kind of human resource is necessary to ensure that the smart city is localized to the features and needs of the locality in question. That is why we have identified human resource development as a key factor. We launched a human resource development program through which we attempt to answer these questions.

Smart-city initiatives, being based on urban planning and information science, cover disciplines such as healthcare and welfare, mobility, logistics, and community design. Therefore, an interdisciplinary team is needed, along with a program to train leaders who can coordinate between the disciplines and unify the team. The human resource development will also need exit strategies so that the leaders trained under the system can go on to serve in public-sector positions where they will lead cross-sectoral efforts or manage smart-city initiatives combining multiple disciplines.

For better or worse, people increasingly get their information from social media, which can often include false information. This trend makes it all the more important to disseminate accurate information about the IT technologies and datasets used in smart cities, improving the public's literacy in smart cities. Chapter 9 presents our research findings on this key factor.

4.7   Key Factor 6: Data Ecosystem

Once smart-city projects get underway, data will begin to accumulate. Project-specific datasets will obviously be valuable for the projects in question, but their value increases exponentially once the data cross-pollinates across different projects, and such cross-pollination could create opportunities for creating new projects and industries. The new businesses created will prove valuable resources when it comes to the sustainable management of a smart city. Thus, critical to the sustainability of a smart city are data cross-pollination and active participation by the private sector.

However, with the datasets simply laid out in the city operating system (as in Society OS) or data-linkage platform, scarcely anyone ever uses the data across different sectors. That is why we have identified having a data ecosystem built by the stakeholders who use the data-linkage platform as a key factor and have researched a cyclical process for creating such an ecosystem.

Data cross-pollination is prevented by the lack of rules concerning the use of data, the distribution of the benefits from the use of data, and the lack of incentive to provide the data. Measures must be taken to address these issues.

Innovation incubators are needed to ensure horizontal communication between enterprises, universities, and the organizations coordinating the community-development projects, and to nurture both the organizations and human resource (coordinators) to facilitate data-linkage and the businesses that will use the data. Chapter 10 presents our research findings on this key factor.

4.8   Applying the Key Factors in a Target Community

At H-UTokyo Lab, we consulted with the national and local governments driving forward the smart-city agenda, and they confirmed that issues pertaining to the six key factors exist in local communities. If there are communities that for want of signposting and directions are lost and unsure how to proceed with delivering a Society 5.0-embodying smart city, then a good place to start is to work out how the six factors should be incorporated into the community in question.

When it comes to disseminating the key factors in cities and communities, the districts and municipalities have an important role to play, but so do the prefectural and national governments, which can provide assistance over a much wider area. In May 2021, we presented the national government with 15 policy recommendations for delivering sustainable smart cities; these recommendations were linked to the key factor (H-UTokyo Lab 2021). In 2023, we published our research findings on the six key factors (H-UTokyo Lab 2023). For national or local government employees involved in smart cities, we hope that these publications offer useful insights for future policymaking.

This chapter outlined how the key factors complementing the Society 5.0 Reference Architecture relate to problems/needs common to all communities and how they point to the solutions to these issues. The purpose in applying the six factors is to ensure that smart cities align with the people-centric ethos of Society 5.0 and are sustainable (so that they do not end up a passing fad).

Put another way, the chapter introduced the key factors for engaging in people-centric smart cities and ensuring that they are sustainable. The next chapters introduce the practical research we conducted into the six factors between 2020 and 2023. Many of the research activities were linked to specific initiatives in cities and communities, so we hope the findings offer valuable insights in how to apply the factors in real-world cities and communities.

References

H-UTokyo Lab (2021) *Teigensho "Jizoku kannō na sumāto shiti no jitsugen ni muketa teigen: 5 tsu no kī fakutā to kuni ni yoru 15 no shiensaku no teian"* [Proposal Toward Realizing a Sustainable Smart City – 5 key factors and our proposal on 15 national support measures-] May 14, 2021. http://www.ht-lab.ducr.u-tokyo.ac.jp/2021/09/07/030/. Accessed March 21, 2024.

H-UTokyo Lab (2023) *"Society 5.0 no ākitekuchā: Hito chūshin to jizoku kannō no ryōritsu"* [Society 5.0 architecture: Balancing people centrism with sustainability] (webinar for Habitat Innovation project, March 17, 2023). http://www.ht-lab.ducr.u-tokyo.ac.jp/2023/03/17/news047/. Accessed March 21, 2024.

Chapter 5
Social Acceptance

Kaori Karasawa

Abstract The chapter analyzes the concept of "social acceptance" in the context of technology and service implementation, particularly in smart cities. Findings from two research projects pertaining to the relationship between people and smart-city initiatives are focused on, and the societal values and goals underlying technological implementation are examined. The first approach highlights individual attitudes, particularly those regarding the use of personal data in smart-city services, with emphasis on the role of trust between citizens and service providers. The second approach analyzes the dynamics of community acceptance or rejection throughout the service-implementation process, which underscores the importance of stakeholder interactions and may involve conflicts in some scenarios. A case study from the Sidewalk Toronto project is referenced, where challenges in garnering social acceptance ultimately resulted in the project's suspension. The chapter concludes by suggesting a reevaluation into the conceptualization of social acceptance, with emphasis on the necessity to consider societal values, the ethical responsibilities of implementing technology, and the broader impact of technology on communities. Addressing these factors is crucial to realizing human-centered smart cities and fostering a harmonious relationship among people, society, and technology.

Keywords Social acceptance · Public attitudes toward the technology · Trust · Trust-based attitude model · Relationship between scientific and technological innovation and society

K. Karasawa (✉)
Department of Social Psychology, Graduate School of Humanities and Sociology, The University of Tokyo, Tokyo, Japan
e-mail: karasawa@l.u-tokyo.ac.jp

© The Author(s) 2025
Hitachi-UTokyo Laboratory (H-UTokyo Lab.), *The Architecture of "Society 5.0"*,
https://doi.org/10.1007/978-981-96-2929-9_5

5.1   Thinking About Social Acceptance

5.1.1   Caring About Social Acceptance

Scientific and technological innovations, along with the services stemming from them, have the power to transform individuals and communities; thus, a number of issues must be addressed when delivering innovations. One of these is social acceptance, which is the topic of this chapter. We identify social acceptance as a key factor in smart-city delivery.

Those in charge of delivering smart cities should care about and value social acceptance, because public acceptance of technology and services is key to facilitating the smooth operation of technology-driven services and increasing the chances of technology delivering desirable outcomes. A smart city should use scientific and technological innovations to deliver convenient and efficient services to the public and address social issues, thereby contributing to well-being. However, people will never enjoy these benefits if they reject technology and services outright or if they feel reluctant to use them. To ensure that innovations can be translated into practical services of benefit, it is necessary to consider people's attitudes toward science and technology, and ways to provide accepting services.

However, this is not the only reason for social acceptance. The immediate goal of delivering scientific and technological innovations is to make life more fulfilling and easier and address the problems people face, but the ultimate goal is to create a more desirable society and a better world. The history of science and technology is littered with examples of unforeseen risks and grave socio-environmental outcomes. To ensure that scientific and technological innovations continue to create new value without inflicting severe damage on individuals and communities, those involved in R&D or delivering its outcomes must identify or foresee the potential harm the innovations could create. Social acceptance is a key concept in examining these potential harms. Social acceptance encourages one to pause and think about what people's needs are and what kind of society is desirable, and to rethink the role of science and technology. It also provides a perspective for reflecting on the relationship between people and science: Who are science and technology for, and who is responsible for governing science and technology?

Amid the complex interplay between society and science, the only way to empower the pursuit of happiness, address socio-environmental problems, and forge a brighter tomorrow in accordance with the people-centric ethos of Society 5.0 is to respect people's values and have a healthy, constructive discussion about where science is taking us. When delivering smart cities, the question of how to approach the challenge of social acceptance is a critical part of the conversation.

5.1.2 How to Approach Social Acceptance

The question then is how exactly do we approach the challenge of social acceptance? What issues should be considered when discussing social acceptance? In this chapter, I present two approaches to this topic and identify some points that can lead to a more fruitful conversation on social acceptance. First, the general content to be discussed is outlined.

In any discourse on social acceptance, the underlying motive is to encourage the public to accept the rollout of technological innovation. The goal is for the public to see technology as a good thing, agree with its implementation, and use the service. One approach is to remove any obstacles to this goal and devise a strategy to facilitate smooth delivery of technology. For example, one might want to organize a PR campaign highlighting the benefits of the service or introduce incentives. One could also address the risks of using technology to address public anxiety. This approach is based on the idea that social acceptance is a matter of individual attitudes and seeks to identify effective measures to help people develop more positive attitudes toward science and technology. We call this the "individual-attitudes" approach.

This approach will likely attract the most attention during the initial and middle stages of the key factors (these stages are discussed in Chap. 14). Smart-city initiatives can scarcely progress when the public feels uneasy about the risks of the technology, when they feel reluctant to use the service, and when the initiatives create suspicion about business operators. Acting swiftly is necessary to foster a more accepting public attitude. We need to understand what the public thinks and act to mitigate negative attitudes, anxieties, and resistance.

However, when taking this approach, one must be mindful of falling into an activity trap in which the means fostering supportive, or at least non-oppositional, attitudes become an end in itself. Although ignoring public opinion would contravene the people-centric ethos of Society 5.0, when endearing the public is treated as an end in itself, one may lose sight of the very reason for doing so and become absorbed when devising strategies for shaping attitudes and crafting persuasive PR campaigns.

Another point to note is that the social acceptance of smart cities is not just a matter of what individuals think; the target community as a whole must be considered. Delivering and rolling out technology will always take some time and will involve interactions and, often, disputes between different stakeholders. This aspect of social acceptance requires attention too; social acceptance is also about how the community's acceptance (or rejection) is formed following a series of processes in which different stakeholders influence each other and debate the benefits and problems of a technology or service. We can call this approach to social acceptance the "interactions" approach.

Using this approach, real-world cases are explored to determine what communications and debates arise between stakeholders, and to identify the contentious issues or talking points that are likely to arise. Through the theme of social acceptance, we consider the issues and contentions that arise concerning the technologies and services to be delivered in a smart-city project.

The next section presents research on the two approaches to familiarize readers with them and engender a more holistic understanding of social acceptance.

The final section offers fresh ideas for encouraging a fruitful discourse on social acceptance. If a pursuit of social acceptance is to contribute toward a people-centric smart city, then we have to understand what "social acceptance" means in the first place. To answer this question, we begin by focusing on the benefits of smart cities and the actors involved.

5.2   Case Studies on Social Acceptance

5.2.1   Individual-Attitudes Approach

This section introduces cases concerning the first of the two approaches, the individual-attitude approach, to illustrate ways of improving social acceptance. At the H-UTokyo Lab, we drew from the literature on social acceptance and added fresh research insights, focusing on trust. This section first summarizes our opinions of the literature and presents our own research findings.

To reiterate, when dealing with social acceptance, the first issue to address is public attitude toward the technology in question or the service stemming from the technology. It is a commonly shared view that the understanding and consent of people in a community are needed to implement measures that affect that community. The same is true for the implementation of smart-city initiatives. Thus, the attitudes that people hold and exactly what underlies them must be identified.

One simple way to assess the level of acceptance of a technology or service would be to ask each person whether he or she accepts or agrees with the technology or service in question, and see what percentage are "in favor" of the answer.

However, according to the findings on social attitudes in the social psychology literature, attitudes might be more complex in reality. According to the literature, attitudes should be understood not as a superficial for/against binary, but as the product of complex associative networks featuring an interplay between cognitions, emotions, and behavioral intentions (Eagly and Chaiken 1993). With this understanding, we can appreciate how the real meaning of each "in favor" response will differ depending on the knowledge, beliefs, values, and emotions at play in each case. This understanding also offers a practical benefit in that it can offer insights into how to make people more favorable.

From this more nuanced perspective, social psychologists have considered multiple facets of perceptions about technology and services and explored a full spectrum of factors that predict acceptance (or rejection) of a technology and service, or intention to use a service. The leading models for explaining these factors are the trust, confidence, and cooperation (TCC model) (Earle and Siegrist 2008) and the technology acceptance model (TAM) (Davis 1989). The literature covers a range of technologies and communities and has empirically demonstrated, mainly using survey data, the roles of variables in determining social acceptance. The key variables have varying effects; however, the acceptance or rejection of technology is generally determined by variables related to the perceived quality of the technology or service and their perceived effects, including whether the risk is low relative to the benefits, whether the technology is designed to meet public needs (or address a problem), whether the implementation costs are low, whether the technology is urgently needed, and whether the technology will not threaten the environment or health.

In view of these insights, to gauge a particular community's social acceptance, one could conduct a questionnaire survey in which residents are asked about their attitudes toward technologies or services, whether they support their implementation, and whether they would use them. Analysis of the response data identified correlations between variables. This analysis provides insights into the attitudes of people at this stage, the factors impeding the acceptance of technologies and services, and the steps that need to be taken to improve them.

5.2.2 Importance of Trust

When a new science or technology is implemented in a particular community, social acceptance of the technology or service may be impeded by conflicts and confrontations between proponents (local governments or enterprises) and residents. One reason for this is a lack of trust in proponents (e.g., Soma and Haggett 2015). It is necessary to promote technological sophistication and appeal to citizens about the benefits of services, but the question is not only what kind of relationship the proponents have established with the residents and will continue to establish in the future.

Trust is a critical factor in determining the acceptance of many kinds of strategic interventions, not just the delivery of scientific and technological innovations; it is also a key concept for the smooth operation of any social system (Koyama 2018). Whether the interventions are by national or local governments, private companies, or other types of societal interventions, the actors try to end the public intervention by sending a message that emphasizes its benefits and why it is needed. However, without trust between the intervening party and the public, these attempts will prove ineffective and increase public misgiving.

The importance of trust becomes clearer when the mental processes that reinforce misgiving and resistance are considered. If we feel that someone is untrustworthy or shifty, then we will never feel reassured by their explanations of risks, benefits, and safety. We also begin to question the intentions behind the measures taken during the implementation process. Officials may insist that the intention is to improve the quality of public services, but we may doubt the honesty of their intentions and even suspect an ulterior motive: the real purpose is to obtain financial profit or to survey and control the public. We would doubt their promise to operate the technology properly and suspect that they would use it for purposes other than what was stated. Thus, although communication with the public is necessary to enhance social acceptance, healthy rapport is not possible without a foundation of trust.

Trust is a decisive factor in determining attitudes toward smart-city technologies and services. This is especially true for smart-city technologies that use personal data. Growing public concerns about privacy make the consideration of social acceptance and the role of trust extremely urgent. In the 2021 study, we conducted a survey in the areas of credit evaluation, AI cameras, and health and welfare, asking open-ended questions about concerns about the provision of services using personal information. We then subjected the responses to text analysis, which revealed several key terms, including misuse, management, privacy, personal identifiability, and supervision (Shimizu et al. 2021). Smart cities rely on the use of such data, but people are mindful of the risk that the data will be mishandled (resulting in data leakage or similar events) or that their individual autonomy would be threatened. If trust shapes such attitudes, an attitude model based on trust must be envisaged when devising ways to enhance social acceptance.

5.2.3   Trust-Based Attitude Model for Social Acceptance

With this problem awareness, we previously researched a case involving a smart-city initiative that used personal data to empirically verify the determinants of trust in a technology (Hashimoto et al. 2020). Figure 5.1 displays the model derived from this study. Trust is the starting point in this model. Trust influences critical acceptance of smart-city initiatives. More precisely, it influences attitudes toward personal data risks, which in turn influence the willingness to provide personal data and acceptance of the service.

Further back in the causal chain, trust was predicted by perceived sincerity, competence, and value similarity. Of these three variables, those pertaining to sincerity and competence have been identified as fundamental cognitive dimensions in group

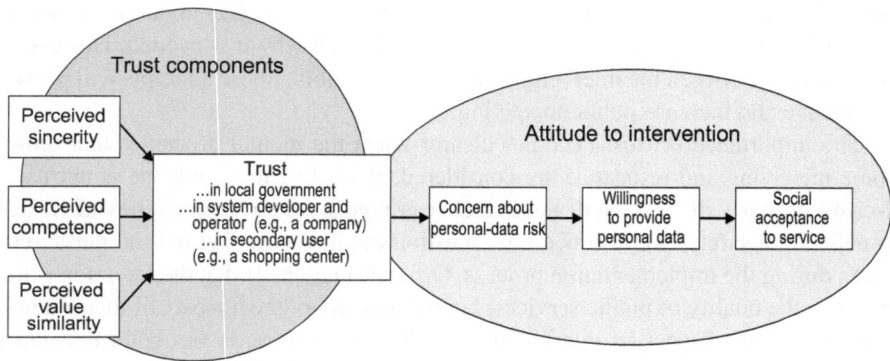

Fig. 5.1 Trust-based attitude model for social acceptance

perception research in social psychology. It has been argued that how groups are perceived in both dimensions determines how people treat them (Fiske et al. 2002). Research applying this finding is also being conducted in the area of risk management (Earle 2010). The third variable, perceived value similarity, is derived from the salient value similarity model, which postulates that people are more likely to trust risk managers who hold similar values to them (Nakayachi and Cvetkovich 2008). Salient value similarity has generally been considered in the context of managing health and environmental risks; however, in our study, we assumed that it is also relevant to the use of personal data in smart-city initiatives.

5.2.4 Empirical Evidence for the Importance of Trust

The online study conducted in September 2019 is described as follows. The respondents were asked to describe how they felt about a hypothetical scenario involving service provision. The sample consisted of 307 men and 318 women with a mean age of 45.8 (with a standard deviation of 14.4). The respondents described the scenario as follows:

> The service will collect foot-flow analytics data from AI-driven cameras. It will also collect data voluntarily from people's smartphones, including GPS data, data about the public transport services the person has been using, and the amount of money the person has been spending. The data will be used to guide the design of shopping districts, the planning of transport services, and the delivery of public services. The service will also allow the participation of secondary users, which may include shopping centers to refer to and use the data.

The respondents were then presented with questions (these questions will be discussed later) pertaining to the trust components in Fig. 5.1: their trust in the system developer and operator of the service (a private company), in the local government that spearheaded the initiative, and in the secondary users of the collected data. They were then asked three questions about their attitudes toward the service: their perception of the risks associated with the use of personal data (e.g., I fear that the data would be misused), their willingness to provide personal data for the service (e.g., I would not mind providing my personal data), and their acceptance of the service (e.g., I think the service is a good idea).

The variables affecting trust in the intervention were subjected to a pathway analysis. Figure 5.2 shows the results. Arrows indicate statistically significant inter-variable pathways. Pathway analysis revealed that trust in actors predicted attitudes toward the service. Although the strength of this association varies depending on the actor, the general trend is that stronger trust is associated with lower concern about risks, a greater willingness to provide personal data, and a more accepting attitude toward the service. These results corroborate the validity of the model in which trust in service proponents is a fundamental variable in determining acceptance.

Figure 5.3 shows a model describing how trust is shaped by the perceived sincerity of the actor (local government, operator, or secondary user), perceived

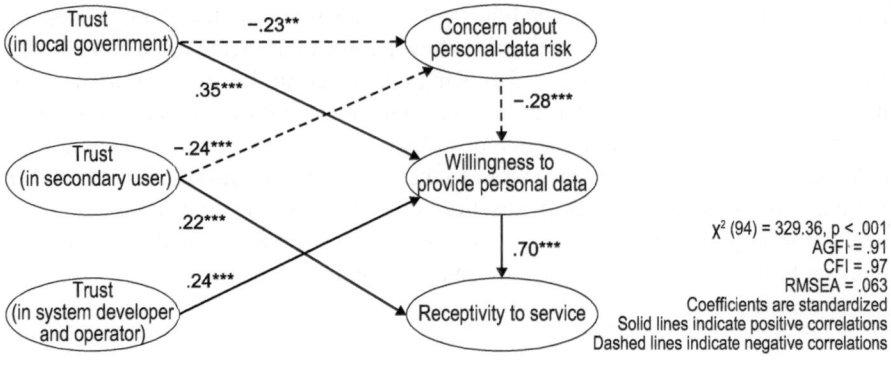

Fig. 5.2 Results of analysis of determinants of trust

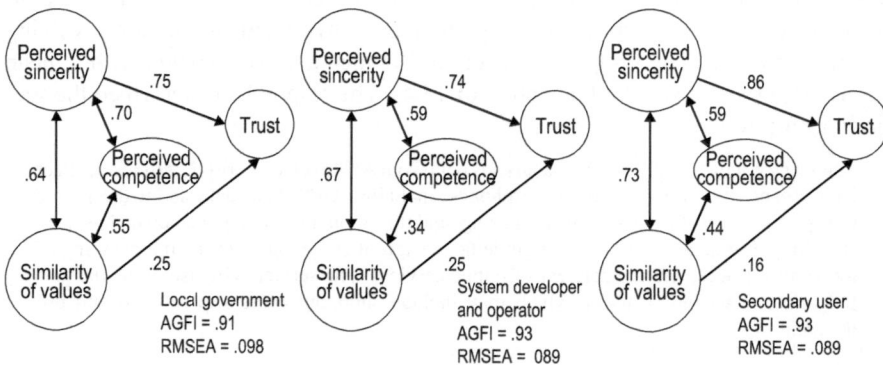

Fig. 5.3 Model describing determinants of trust in local government, operator, and secondary user

competence of the actor, and similarity of the actor's values. Regardless of the actor, perceived sincerity and similarity of values are associated with stronger trust. Perceived competence had no direct pathway to trust, but was correlated with perceived sincerity and similarity of values.

5.2.5 How to Gain Trust

This study may have demonstrated the importance of trust and revealed the factors that predict it, but this leads to another question: How, then, do we ultimately gain trust? Trust is not built overnight. We also know from everyday experiences that problematic behavior can destroy trust after a stroke. Can we specify a clear-cut, fool-proof technique to gain trust? Although this may not be possible, our results do at least offer some clues. The implications of this study are as follows.

What we need to know here is which of the question items used in the above study (and other related studies we conducted at the H-UTokyo Lab) predicts the person's perception of the actor's sincerity and competence. In a series of explorative analyses, we set up a number of questions to determine which appeared to be correlated. The analyses revealed that perceived sincerity correlates strongly with agreement in the following statements: "the actor is transparent about the service," "the actor will sincerely handle any troubles that may occur," "the actor treats all residents equally," and "the actor makes fair decisions." These statements describe a person's perception of how the actor engages with the residents in the process of delivering services. As such, the results imply that actors can gain trust by showing sincerity and impartiality in their everyday dealings with citizens. The analyses also revealed that perceived competence correlates with agreement in the following statements: "the actor has expert knowledge and skills," "the actor is competent using the latest technology," "the actor's service is of a high quality." These results suggest that a positive appraisal of an actor's knowledge and technology engenders confidence in the actor's ability to be given responsibility, and this confidence strengthens perceptions of sincerity and value similarity.

We also found that value similarity correlates with agreement in the statement "the actor values what I value." This result suggests that actors, while gauging public opinion, should also try to create emotional resonance by emphasizing how the service will address socioeconomic problems or needs/values to be fulfilled.

The exact trust-gaining strategies that are available and effective will always depend on the circumstances of the case in question, including the system available at the frontline workplace and the talent deployed in frontline roles that interface with the public. Still, the study findings do suggest that, regardless of the case, much will hinge upon the way the individuals concerned conduct themselves when engaging with the public (such as conducting oneself in a way that creates the impression that "the actor sincerely handles any problem that may occur"). If we make this just a matter for individuals (employees or other individuals who directly engage with the public), then it may obscure an improvement procedure. It is important to consider organizational actions to rectify issues in frontline workplaces, such as testing workplace design, providing training to improve awareness in the workplace, or introducing a feedback system. In this way, the management of the workplace environment can be another route to gaining trust.

5.2.6 The Interactions Approach

Next, we consider examples of studies that use the interaction approach. This approach frames social acceptance as a product of interactions between stakeholders. I have explained why this is important, but it is worth going into more depth before presenting my research findings.

Not all residents are familiar with the services rolled out in a smart city. In January 2022, we polled a thousand Sapporo residents about whether they were aware of certain services introduced to promote health and improve transportation convenience (including *Kenkō Point*, *Jinryū Center*, *Satsu Navi*). The results varied by service; however, the awareness was generally low, ranging from 3% to 5%. Given this trend, attitudes toward a particular service may be ill-informed. As such, attitudes may be easily swayed by a person's intuitive impressions of the actor, word of mouth, and media.

We must also understand that a community's acceptance of a service is not simply the sum of the attitudes of each member of the community. Rather, it is shaped progressively by a series of events occurring along a time series, including engagement by the actors (local government or private companies) in the process of delivery (which could include communicating information, holding briefing sessions, or responding to residents' concerns), media reporting, activism by the pro- and anti-camps, and, more recently, discussion sessions in living labs and workshops.

That is, the attitudes held by individuals offer no insight. These attitudes provide a snapshot of what the public perceives and understands at a given time in a sequence of events. This information has proven important in the discourse on social acceptance and has been analyzed using the individual-attitude approach. However, to understand how a community's acceptance (or rejection) of a technology or service develops over time, one needs to trace its development across a sequence of chronological events and clarify the processes by which stakeholders interact with one another. Such an analysis can also help consider the issues that social acceptance involves and learn from past cases. One such case is that of Sidewalk Toronto, an initiative in Toronto, Canada.

5.2.7 Opposition to Sidewalk Toronto: Background and Talking Points

Sidewalk Toronto was a project designed to redevelop an area along Toronto's waterfront. The plan was shelved following public pushback. To research this case, we collected open-access information pertaining to the interactions between key stakeholders: the local government, businesses, citizens, and local media. We then arranged the information in a time series and identified the problems that occurred.

The synopsis is as follows: The plan stemmed from a government agenda that was returned in 1999. In 2017, Sidewalk Labs, a subsidiary of Google's parent company Alphabet, began engaging in the planning process. The result was the launch of Sidewalk Toronto and the establishment of an organizational framework for delivering the project. Following the launch, the project team consulted the public and solicited feedback through several town meetings and public roundtable discussions. Nonetheless, problems and concerns regarding the project were raised, which led to growing opposition by civic groups. The Canadian Civil Liberties Association filed a lawsuit against the project, prompting calls to shelve it. In May 2020, Sidewalks Labs announced that it would withdraw from the project, citing the economic uncertainties created by the pandemic.

We analyzed this chronology and identified the sources of contention that led to public opposition. People cited misgivings about data governance (particularly that the data would be stored abroad) and that the project was driven by a private company. People have also complained about this process. The project team organized briefing events, but the briefings focused on the innovative aspects of the project, fostering the impression that the project team was prevaricating on the public's main concern—how personal data would be handled (e.g., whether privacy would be violated or whether the data would be used for surveillance)—and giving rise to the suspicion that the project was just a profit-making scheme for Sidewalk Labs. The organization in charge of addressing concerns regarding data privacy was the Digital Strategy Advisory Panel, an independent advisory panel for Waterfront Toronto (a government agency). A member of the panel resigned in disgust, claiming that Waterfront Toronto had failed to address public concerns over the data. Meanwhile, the public was becoming increasingly alarmed about a proposal that would guarantee Sidewalk Labs' profits. These misgivings, combined with increasing skepticism about Google's role behind the curtains, amplified the pushback.

Notably, public pushback was provoked not so much by the service itself, but rather by contention over principles such as civil liberties (privacy) and democracy. The pushback came to the head when a group of citizens launched the Block Sidewalk campaign, with a message on their website affirming that "democracy is not for sale."

I discussed the importance of an individual-attitude approach earlier, but the case of Sidewalk Toronto is an example of a failure to build a foundation for trust. Lack of trust came to light around the issues of data governance and corporate interests, and in an atmosphere of doubt, people's attention reverted to more fundamental social concepts such as civil liberties, responsibilities, community autonomy, and democracy. The lesson from Sidewalk Toronto is that the challenge of social acceptance requires us to reexamine how technology and services relate to individuals and society.

5.3 For a Fruitful Conversation About Social Acceptance

5.3.1 Rethinking What Social Acceptance Means

In discussing the two approaches to social acceptance, I emphasize the importance of trust in winning social acceptance and present evidence illustrating how controversy can be invoked with emotionally charged language pertaining to fundamental social values (such as rights, democracy, and responsibility). However, this does not imply that trust should be gained purely for the purpose of enhancing social acceptance or that one should try to steer clear of any loaded or emotive language, for that would confuse ends with means. If the goal of enhancing social acceptance has transfixed us, we might end up simply chasing public opinion without seriously considering the type of future being created for society, the role of science and technology in this future, and how a people-centric society powered by science and technology will deliver well-being. The challenge is that we must still endeavor to win social acceptance but must also keep exploring ways to avoid the trap of enhancing social acceptance for its own sake.

To address this challenge, the meaning of social acceptance must be reconsidered. What does social acceptance mean in the context of accomplishing the values we resort to, and what do we mean when we mention "social acceptance" in a discussion about building a better society? (Todayama and Karasawa 2019). This analysis will require a critical evaluation of the existing discourse on social acceptance.

When people debate ideas for a better society, they use language, meaning that talking points and agenda-setting are always constrained by semantics. This has also been true for social acceptance: whenever "social acceptance" was used in the sense of individuals holding positive attitudes toward the implementation of the technology or service in question, the discourse was steered along the individual-attitudes approach. Whenever the term was used in the sense of the community as a whole, accepting the implementation, the discourse was steered along the interaction approach. When the discourse was steered along the individual-attitudes approach, "social acceptance" would usually connote the factors necessary for people to form favorable attitudes. When discourse is steered along the interaction approach, "social acceptance" usually emphasizes the talking points that ought to be managed to prevent community pushback. In both cases, the underlying agenda was to motivate individuals or the community as a whole to adopt a more accepting attitude toward the scientific and technological innovation in question so that the benefits of the innovation could be realized. This binary relation suggests that the discourse about social acceptance was being driven by a "benign" motive to deliver the benefits of scientific and technological innovation to society and, as such, would gravitate to the matter of how to make individuals or the community more accepting.

To stop the discourse on social acceptance from becoming locked into this single agenda, we need to return to the basics and relate social acceptance to the fundamental questions of what we want to achieve by building smart cities and what kind of society we want to achieve. With the advancement of science and technology, we need to discuss how individuals and communities relate to science and technology and what choices for the future we should be making. Perhaps a useful way to contribute to a more fruitful debate on these questions is to work out on how "social acceptance" should be defined and what fundamental issues the discourse on social acceptance should be focusing on.

This is not an easy task and there are no easy answers. However, once we take the discourse on social acceptance back to the original question of how people and communities relate to science and technology, and then ask questions such as what options are available to people and communities and who exercises agency in these decisions, we will get a step closer to the heart of the issue. In the concluding section, I discuss these points and offer conclusions on what the discourse on social acceptance should focus on and how it should proceed in the future.

5.3.2 Values a Smart City Embodies and the Actors

The purpose of technology is to produce something valuable to society. When technology is implemented, the society enjoys value. The value might be greater convenience, efficiency, and comfort or it might be a solution to socioeconomic needs, such as an anti-crime solution, a wellness solution, or a solution to an environmental problem. Social acceptance, then, can be interpreted as endorsement (if the people are "socially accepting" a value, it means that they endorse the value). However, what is the value gained by implementing a neutral and transparent technology? Is technology equally valuable to everyone? Does everyone benefit? Is the nature of values clear to everyone?

In many cases, the value generated by technology requires financial payment or literacy. Rich, scientifically and technologically literate people are more likely to benefit than others. The degree to which a person benefits may also be shaped by the size and economic clout of the community to which they belong, or by demographics. There is no guarantee that technology-embedded value offers neutrality or transparency, meaning that disparities can arise among those who enjoy the value. To clarify, I am not saying that it is inherently bad for technologies or services to be payable or require literacy. The problem I highlight here is that the nature of the embedded value can be obscure, causing people to disregard questions about its neutrality and transparency.

Once we put aside the premise that technologies and services offer neutral and transparent value, the next question is who decides what is valuable and who benefits. This is really a question of what social entity we have in mind when we say "social" acceptance, or who are the actors exercising agency in a smart city. "People" or "citizens," some might say, but who are the actors in reality? If a smart city is an opportunity to shepherd people's attitudes and behaviors (and the direction of the social entity as a whole) in a more desirable direction, who is shepherding and in what direction are people being herded?

The theme of social acceptance can prove useful in identifying the key issues related to these questions. One such issue is that emphasis on social acceptance leads us to make choices with which as many people agree as possible. In other words, there is a trap in deciding that a technology or service should be based on the "aggregation of people's attitudes." While it is undeniable that the approval of many people is important, the issue of dealing with disparities, inequalities, and minorities, as well as the fact that the approval of many does not guarantee the "rightness" of the choice itself, must be reflected in discussions on social acceptability.

Perhaps the answer here is to respect the diversity of individuals. Theoretically, this could mean focusing on whether a technology or service satisfies individual users and provides a system that ensures that the technology or service faithfully reflects the particular needs of each user. This implies that one is trying to contribute to the well-being of every individual. Although this might be the case, what accommodating diversity really means, whose needs should be accommodated, whose needs can be disregarded, what the operating principles are, and who decides on them, still needs to be considered. Moreover, although satisfying individual needs and desires may seem preferable, should we not be wary of stifling personal growth or fostering a mindset that never questions the relentless pursuit of desire?

Regarding the second issue, the aforementioned problem of equating social acceptance with a consensus of opinion intersects with the problem of treating social acceptance as a process that alters people and society. If the goal is to alter people and society such that they can accept a technology or service, then the goal can never be people-centric. If altering people and society is wrong, then perhaps we should take the opposite approach to see where the majority opinion lies, and then develop technology and services reflecting this opinion. To some extent, such an approach would be welcome because it respects people's attitudes. However, this study has some limitations. When decisions are shaped by social acceptance, people may feel more satisfied with the outcomes; however, is this always a good thing? If we worked to give everyone exactly what they wanted, we would force the next generation to pick up the bill. If we proceed according to the reasoning that we have developed only the technologies and services that people accept, then it would become uncertain where responsibility lies, creating a deficiency in the governance of science and technology.

Presented with these issues, the impression is that social acceptance is fraught with complex and bothersome matters. These issues must be addressed when delivering scientific and technological innovations to society. Social acceptance can potentially help develop discussions on these issues, depending on the sense in which the term is used.

When considering the semantics of social acceptance, note that based on how the term is used, it may be necessary to consider the possibility of a shift from a passive position, in which people accept the benefits of technology, to one in which they proactively decide what the technology should be. In past discourse on social acceptance of a smart city, the focus was on whether the people living in the city would be "beneficiaries" of the benefits and risks of the technology or service. However, when people are regarded not just as passive beneficiaries of benefits and risks but as active agents who decide on what values to prioritize, then we reach the limits of a dichotomous framework that sees social acceptance in terms of an exchange between the "benefactor" (a local government or private company delivering the service) and the "beneficiaries" (the people receiving the benefits and risks of the service). In one sense, social acceptance means that beneficiaries receive a technology and service based on the technology from the benefactor. However, social acceptance needs to be considered in another sense: the sense that both the benefactor and beneficiaries actively engage in the decision-making process regarding the technology itself and the implementation of the services based on it. Social acceptance then forms the foundation for relationships between the groups of actors who make up a "social" entity—the developers of the technology or service, the people who choose the technology or service, and the people who choose what values to prioritize.

If we consider this perspective, we should aim for a more holistic discourse that considers the processes that influence diverse individuals in society to make important decisions about their futures, the mechanisms that enable such decisions, and the sovereignty and responsibilities of these people as citizens. Such a discourse would call for collaborative engagement in efforts to build a healthy relationship between scientific and technological innovation and society. As part of this, the connotation of social acceptance would shift from the idea of people as passive beneficiaries to people as active agents.

References

Davis, F. D. (1989) Perceived usefulness, perceived ease of use, and user acceptance of information technology, Management Information System Quarterly, Vol.13 (3): 319–340, September 1989, DOI:https://doi.org/10.2307/249008

Eagly, A. H. & Chaiken, S. (1993) The Psychology of Attitudes, Fort Worth, TX: Harcourt Brace Jovanovich College Publishers.

Earle, T. & Siegrist, M. (2008) Trust, confidence and cooperation model: A framework for understanding the relation between trust and risk perception, International Journal of Global Environmental Issues, Vol. 8 (1): 17–29, February 2008, DOI: https://doi.org/10.1504/IJGENVI.2008.017257

Earle, T. C. (2010) Trust in risk management: A model-based review of empirical research, Risk Analysis, Vol. 30 (4): 541–574, April 2010, DOI: https://doi.org/10.1111/j.1539-6924.2010.01398.x

Fiske, S. T., Cuddy, A. J. C., Glick, P. & Xu, J. (2002) A Model of (often mixed) stereotype content: Competence and warmth respectively follow from perceived status and competition, Journal of Personality and Social Psychology, Vol. 82 (6): 878–902, June 2002, DOI: https://doi.org/10.1037/0022-3514.82.6.878

Hashimoto, T., Tham Y-J., Karasawa, K., & Tai, K. (2020) *"Dēta kudōgata shakai" ni taisuru hitobito no taido kōzō* [Construction of people's attitudes about the data-driven society] In: Poster presentations of the 84th conference of the Japanese Psychological Association, Toyo University, Tokyo (online), September 2020.

Koyama, T. (ed) (2018) *Shinrai o kangaeru: Rivaiasan kara jinkō chinō made* [Thinking about trust: From Leviathon to AI]. Keiso Shobo, Tokyo.

Nakayachi, K., & Cvetkovich, G. (2008) *Risuku kanri kikan e no shinrai: SVS moderu to dentōteki shinrai moderu no tōgō* [Trust of risk managers: An integration of the SVS model and the traditional view of trust]. Japanese Journal of Social Psychology Vol. 23(3): 259–268, DOI: https://doi.org/10.14966/jssp.KJ00004896221

Shimizu, Y., Osaki, S., Hashimoto, T. & Karasawa, K. (2021) How do people view various kinds of smart city services? Focus on the acquisition of personal information, Sustainability, Vol. 13 (19), 11062, DOI: https://doi.org/10.3390/su131911062

Soma, K. & Haggett, C. (2015) Enhancing social acceptance in marine governance in Europe, Ocean and Coastal Management, Vol.117: 61–69. November 2015, DOI: https://doi.org/10.1016/j.ocecoaman.2015.11.001

Todayama, K., & Karasawa, K. (2019) *Gainen kōgaku" sengen! Tetsugaku × shinrigaku ni yoru chi no enjiniaringu* [Conceptual engineering declaration: Engineering knowledge by multiplying philosophy by psychology]. The University of Nagoya Press, Nagoya.

Chapter 6
Data Governance

Toshiya Watanabe and Tadashi Mima

Abstract This chapter discusses the concept of data governance in smart cities. There are legal considerations when using data, such as personal information protection law and the protection of trade secrets. However, even if laws and regulations are not violated, this can be a major problem if stakeholders such as residents are not receptive to smart-city services. While digital technology is an important enabling technology for smart cities, it should be recognized that the risk of data leakage is not zero, especially when dealing with personal information and trade secrets. It is important to explain the risks of data collection and use to data providers, and to fully explain the balance of risks and benefits to stakeholders to gain their consent for data use. In this chapter, the basic concept of data governance, following the approach of our Smart-City Data Governance Guidelines, will be explained.

Keywords Data governance · Privacy governance · Agile governance · Data protection · Smart-city data governance guidelines

6.1 Relationship Between Smart Cities and Data

Smart cities are intended as technology- and data-driven initiatives for solving the complex socio-environmental problems plaguing many municipalities and communities across Japan, including problems related to a shrinking and aging population, frequent natural disasters, and—more recently—the need to reduce greenhouse gas

T. Watanabe (✉)
Research and Innovation Office, The Institute of Science Tokyo, Tokyo, Japan

The University of Tokyo, Tokyo, Japan
e-mail: watanabe.t.7d88@m.isct.ac.jp

T. Mima
Smart Infrastructure Consulting Department, Hitachi Consulting, Tokyo, Japan

Graduate School of Media and Governance, Keio University, Kanagawa, Japan
e-mail: tmima@hitachiconsulting.co.jp

© The Author(s) 2025

Hitachi-UTokyo Laboratory (H-UTokyo Lab.), *The Architecture of "Society 5.0"*,
https://doi.org/10.1007/978-981-96-2929-9_6

emissions and control the risk of infectious diseases. Digitally driven initiatives often require data and AI. Consider, for example, a municipality with an aging population. One issue this municipality faces is that when elderly residents venture out alone, they sometimes go missing or suffer a mishap. When an elderly resident goes missing, the person's family members might contact the local government. To safeguard the person as soon as possible, the local government will use an emergency radio system to issue public broadcasts asking the public to contact them if they see, for example, someone wandering around in out-of-season clothing, sleepwear, or indoor slippers.

If the municipality has a closed-circuit television (CCTV) system, then the task would be easier; the authorities could use an AI program to analyze the footage for a person matching the missing person's description. What we today refer to as AI is a data-dependent technology; it is based on machine learning, a technology that uses a computer program and analyzes big data to infer rules and patterns. As such, AI requires first of all a ground-truth dataset to train the algorithm in the relevant attributes. Having developed an inferential model from the ground-truth dataset, the algorithm can scan human images in the CCTV footage and identify a human image with the relevant attributes.

Data and AI applications go beyond analyzing the CCTV footage. They are used in a plethora of smart-city services. Consider these examples of initiatives being piloted. One type of initiative involves transportation and mobility services that make use of data about the public transport services people happen to be using at the time or data about transportation patterns nearby. Another type involves using real-time disaster outbreak information during emergencies. With the increasing global demand to restrict greenhouse gas emissions, it is necessary to have data about the community's electricity demand and its supply. The same goes for risk of infectious disease; when the Covid-19 pandemic emerged in 2019, governments around the world used apps that utilized data on human body contact and other data-driven measures to encourage social distancing and minimize infection risks.

Smart cities must be data-driven too, and by using data and AI they can deliver useful services to residents. In conclusion, what makes smart cities valuable to communities is the services that address the community's complex socio-environmental problems, and AI and data play an indispensable role in improving the delivery of these services.

6.2   Risks in Using Data and AI

Although data and AI are indispensable to smart cities and enable the benefits of smart cities to be delivered to residents, they are not without their risks. To understand these risks, we must first understand the nature of the data used in smart cities.

"Data" is not a one-size-fits-all term. The nature of data varies depending on the case. Healthcare services, for example, use personal data. Personal data feature in many smart-city initiatives, but this is not the only type of data to do so. To supply a city with renewable energy (such as solar power), business operators' data on their electricity demand are collected. Such data differ from personal data because they come not from individuals but from the organizations (such as private companies) demanding electricity.

In disaster management, environmental data, including precipitation, wind velocity, river levels, climatic conditions, civil engineering works, and transportation data, are used. To see whether residents have access to a means of mobility, congestion data could be used. Although congestion data stem from people's activities, they still differ from personal data.

Thus, the data used in smart cities include both personal data and impersonal data. Impersonal data can be further categorized into corporate data (data pertaining to an organization's operations) and environmental data. These three categories of data differ widely in terms of the type and nature of risk associated with their use. Thus, each of them requires a different set of rules for their use and management approaches.

AI comes with its own risks. However, as we define AI as a data-dependent machine-learning process, many of these risks are deeply connected with data, in that they stem from the data the AI uses.

The government has shown considerable interest in how to govern the risks of AI (Ministry of Economy, Trade and Industry 2022), but we shall leave the risk governance aside for the time being. Presented below are examples of risks associated with the three categories of data—personal data and the two types of impersonal data.

1. *Examples of Risks Associated with Using Personal Data*

In September 25, 2015, the UN General Assembly adopted the 2030 Agenda for Sustainable Development, setting the 17 Sustainable Development Goals (SDGs) (Ministry of Foreign Affairs of Japan 2023) and pledging that no one would be left behind ("we wish to see the Goals and targets met for all nations and peoples and for all segments of society"). The principle here was that in addressing global problems such as poverty, UN members would ensure that no one, including the poorest and most vulnerable, would be left behind. This same principle should also be adopted for smart cities.

To ensure that smart cities benefit all residents, we must always understand what residents need. To that end, their personal data is needed. However, personal data must be handled with care, as it consists of data that could be used to identify the person, such as the person's name, date of birth, address, and photograph of their face. If information about survivors, for example, contains such personally identifiable data, any number of people could access the data and use it for nefarious purposes. Someone might use the information to stalk the person, or use it to send nuisance emails or try to scam the person. Accordingly, when a service uses personal data, it is vital that the personal data is properly handled. In Japan, the handling of personal data is regulated strictly under the Act on the Protection of Personal Information (more on this later).

As sacrosanct as personal data may be, the stark reality is that data breaches are increasing year by year as cybercrime grows ever more devious. According to one survey in 2021, 120 companies in Japan reported "data leakage" incidents, 137 reported "data loss" incidents, and these incidents affected the personal data of 5,749,773 people (Tokyo Shoko Research, LTD. 2022). These stark statistics alone suggest that in addition to strictly managing entrusted personal data, businesses should understand that they can never completely eliminate the risk of a data breach. On this assumption, businesses should hold no more personal data than is absolutely necessary.

Even if there is no danger of the data being leaked or misused, people may still be reluctant to provide their personal data because of concerns over how it would be handled.

Take, for example, COCOA, a contact-tracing app Japan used during the pandemic. The app used Bluetooth to log times in which users were in close proximity to one another and to alert them if they were close to someone who later tested positive for Covid-19. The service was discontinued in 2022. The app collected data about the user's infection history and proximity to people who tested positive. It was designed so that this information would never be shared with other users. Nonetheless, a questionnaire revealed that many people were reluctant to download the app for fear that their data would be divulged to the government or their workplace (Watanabe and Hirai 2022). This example illustrates a problem that often occurs with services for which personal data are used: misplaced fears discourage people from providing the data necessary for the service to be delivered.

In another example, in January 2022, the Digital Agency announced a timetable for the digitization of education, which would involve the unified management of personal data related to lifelong learning. This news sparked fears that the government wanted to monitor individuals' lifelong learning records, and the government had to issue clarifications. In reality, the government had no intention of tracking users' educational history: it simply wanted to enable learners to receive guidance and public assistance matched with their learning history and to enable the data to be used for post-education occupational training and lifelong learning. The data were to be stored and managed across decentralized storage platforms. Nonetheless, the plan met with a fierce backlash on social media.

Controversy and pushback over personal data has often occurred in the case of private-sector services that use personal data. Even if there are no legal issues, severe pushback by users can make the business or service untenable. As smart-city services require people to provide their personal data, they also require public acceptance to the idea of providing their data and to the way the data are handled. Conversely, in the case where acceptability toward such data use is notably poor, controversy and pushback are all the more likely to occur, making it all the more difficult to get the public to accept the smart city as a whole.

First and foremost, one must ensure that legal and regulatory requirements are met in data handling. That means obtaining the person's consent about the purpose for using the data and the method by which it will be used. However, consent alone does not solve all problems. Consider the case of CCTVs. A CCTV could capture scores of random passers-by, none of whom expressly consented to appear on film. In this case, the data provider whose data is being observed by the data acquisitor (i.e., the party whose data is being observed by the party that obtains the data) has an important interest in the use of that data. Thus, many interested parties exist upstream in the data chain. In the case of CCTVs, if observees later learn that CCTV footage of themselves has been used, they might object, claiming that their images were used in an undesirable manner without their knowledge. The ensuing outcry could then threaten the continuity of the service itself.

2. *Examples of Risks Associated with Using Corporate Data*

The same kind of risks apply to corporate data—data pertaining to an organization's operations. I already cited the example of electricity demand among companies. The target community in smart-city services have many business operators, each engaging in their own set of operations. The data smart-city services use can include data pertaining to these business and commercial operations. During the pandemic, attempts were made to track footfall and takings in restaurants and retail stores to strike a balance between restricting movement (to control infection risk) and keeping the economy going. The data showed the times at which footfall is highest and how this contributes to sales revenue. The problem here was that this could be information that companies would not want their competitors to know. Similarly, production and distribution businesses have plenty of information they would prefer to conceal from their competitors. Such data would normally be classified as trade secret and held in strict confidence. As in the case of personal data, when a service operator is entrusted with commercially sensitive information, it will need to properly handle the information and in accordance with the terms of an agreement signed with the information owner.

Just as with personal data, if an interested party has any misgivings—even if there are no legal issues—then these will need to be addressed.

Consider a program that uses innovations in AI to retrieve public-domain information about individuals or companies with a relatively high degree of accuracy. Confidence in the service itself could be lost if interested parties feel that the service is displaying or using such information in undesirable ways (even though this information was obtained from the public domain in the first place). Thus, the observee problem—public data being used to observe owners of the data—applies to both corporate data and personal data. Consider another example. Low-orbiting satellites can now obtain video data with accuracy to the tens of centimeters. The technology could stoke fears about its threat to personal privacy and about commercially sensitive information. Regardless of the fact that the video data is in the public sphere, as long as individual or corporate observees have misgivings, it will be hard to deliver services that rely on such data.

3. *Examples of Risks Associated with Using Other Impersonal Data*

Data on environmental conditions, weather, or natural disasters generally avoid the problems of interested parties and observees. However, if the data are mismanaged, a serious problem could occur, especially if AI is being used. As such, environmental data also require data governance.

The problem concerns the authenticity of the information—whether the data is correct. This problem can also apply in the case of personal data and corporate data. If personal or corporate data are inaccurate (if the information has been recorded inaccurately), then the service might cease to operate properly. Suppose, for example, inaccurate information used in disaster management sends a rescue crew to the wrong destination; the ramifications could be grave. The problem is all the greater with datasets used to train AI models. If a dataset contains inaccuracies, the AI will make faulty judgments. As for the AI training dataset, this will have been modified by manual processing, enabling the AI to recognize particular video data. For this process, it is important to manage the data logs so that one can tell who implemented such processing and how the person's behavior was controlled[1] (AI Data Consortium 2022).

Use of AI also entails a moral problem. Consider an AI-driven self-flying drone that navigates using video footage from a camera. The drone, perceiving the situation in the vicinity, might be used for delivering goods or for nature conservation. However, the same drone technology and data could potentially be repurposed as a self-flying anti-personnel weapon. The topographical data the drone uses may seem harmless, but it could potentially be divulged to a foreign organization or syndicate and used for military purposes.

6.3 How to Treat the Risks Associated with Data Use

Having discussed the risks in using data, I will now discuss how to minimize these risks. The first step is to understand the relevant rules and comply with them. In this context, "rules" refers to explicitly codified laws and regulations. In addition to complying with these rules, we must consider moral requirements.

We previously discussed the examples of COCOA, the digitization of education, and examples involving AI applications. None of these examples involved any legal or regulatory breach, yet the businesses in question were threatened by controversy and pushback because of misgivings about the purpose of the using data or the unaccepted outcome. This latter kind of risk is ambiguous; the extent of misgivings and controversy will depend on how the observees and data providers perceive the service, meaning that there are no black-and-white criteria, and it is never completely

---

[1] The AI Data Consortium has examined in depth how logs of data transactions are managed by analyzing contracts for data sharing on the consortium's platform.

AI Data Consortium (2022) Report on Data Provision Contract for the AIDC Platform.

https://aidata.or.jp/wp-content/uploads/2022/02/Report-on-Data-Provision-Contract-for-the-AIDC-Platform.pdf. Accessed March 27, 2024.

clear how residents will take to the service until they start using it. Thus, legal and regulatory compliance alone can never clear away the risks. In addition to following the explicit rules, one must be aware of other requirements. Both the rules and these other requirements are part of data governance.

1. *Avoiding Legal and Regulatory Breaches*

Of the legal and regulatory requirements for personal data, the most important in Japan is the Act on the Protection of Personal Information (Act on the Protection of Personal Information 2003). The purpose of the legislation is to safeguard the rights and interests of individuals while ensuring due consideration for the value of personal data. To that end, it established common rules for national government organizations, incorporated administrative agencies, universities, local public bodies, and also all business operators and organizations who use personal data. The law provides a definition for personal information. It also defines as "sensitive personal information" personal information that requires special care in that it could cause discrimination or other human rights issues[2]. [7] In addition to specifying rules on the storage and management of personal data, the law makes obtaining the person's consent for the information to be shared with a third party and issuing a public notice if ever the personal data are leaked mandatory.

However, in the case of smart cities, other legislation may apply too. Smart-city initiatives might involve local public bodies using personal data related to healthcare or nursing care, and the use of such data might additionally fall under the Act on Assurance of Medical Care for Elderly People, National Health Insurance Act, and the Long-Term Care Insurance Act. Thus, to deliver such services, legal and regulatory requirements that might be involved must be clarified.

When data related to business activities are used in smart-city services, checking whether the data include anything that is protected as part of contractual, legal, or regulatory requirements is necessary. When entrusted with data that is normally managed by a company or other organization, one will have contractual obligations regarding the use of the data. The contract might, for example, limit the ways the data can be used or shared with a third party. Compliance with these contractual obligations is necessary. Failure to do so would incur the risk of a damages claim, and depending on the terms of the contract, an injunction might be taken out to suspend the service as a whole. Suppose that one was entrusted with commercial data that included commercially sensitive information or limited acesss data. If such data were used or shared in violation of the contract, that company might be subject to an injunction (order for prohibition). If an action constitutes an unfair act specified in the Unfair Competition Prevention Act (Unfair Competition Prevention Act 1993), the sanction given would depend on whether the data were protected as trade

---

[2] Information about the person's ethnicity, creed, social status, disease history, and criminal record requires extra care because it could potentially be a cause of unfair discrimination, prejudice, or other mistreatment.

secret[3] (Ministry of Economy, Trade and Industry 2021) (data that the entrusting party manage in strict confidence) or protected as limited access data[4] (Ministry of Economy, Trade and Industry 2021) (data that the entrusting party only share under strict conditions). Thus, to use corporate data in a smart-city service, the first step to avoid risks is to clarify what legal protections the data falls under and then comply with the legal requirements.

Other impersonal data require care too, as they can potentially fall under a considerably wide range of legislation. For example, data related to transportation and mobility could fall under the Road Traffic Act, whereas data related to communications could fall under the Radio Act.

2. *Exceed Legal and Regulatory Requirements as Necessary*

Complying with these legal and regulatory standards represents the bare minimum level of governance. Remember, even if a service never breaches any legal or regulatory requirement in the use of data or AI, the service may still experience public pushback and controversy over such use, thereby threatening its continuity.

If a service is to use personal data, it is necessary to address this risk, especially by giving consideration to privacy. For this, privacy governance is needed, meaning a form of personal data governance in which there is both compliance with legal and regulatory requirements and the requirements are exceeded by addressing privacy concerns. The approach to building privacy governance will depend on the level of social acceptability toward data use and AI. Social acceptability toward the use of data can vary with the development of technology. For example, social acceptability toward a given type of dataset may change if people understand the idea that technological innovation has made obtaining sensitive information from that dataset possible (the information could not have been obtained in the past). Thus, the range of issues to be addressed when using personal data will depend not just on the scope of relevant legislation (such as the Act on the Protection of Personal Information) but also on the information contained in the dataset, the technological trends, and the socio-environmental trends. As such, privacy governance, which considers a broader extent of issues, is needed (Ministry of Internal Affairs Communications & Ministry of Economy, Trade and Industry 2021). In such a case, the people concerned about potential risks and other matters that could potentially affect their interests should first be briefed, and when the service has started to be delivered briefings are provided whenever circumstances change (Fig. 6.1).

---

[3] Criteria for trade secret Ministry of Economy, Trade and Industry (2021) *Eigyō himitsu: Eigyō himitsu o mamori katsuyōsuru* [Trade secrets: Safeguarding and Using Trade Secrets]. https://www.meti.go.jp/policy/economy/chizai/chiteki/trade-secret.html. Accessed March 27, 2024.

[4] Criteria for restricted data: Ministry of Economy, Trade and Industry, Intellectual Property Policy Office (2021) *Gentei teikyō dēta ni tsuite* [On restricted data]. December 20, 2021.
https://www.meti.go.jp/shingikai/mono_info_service/sports_content/pdf/002_s01_00.pdf. Accessed March 27, 2024.

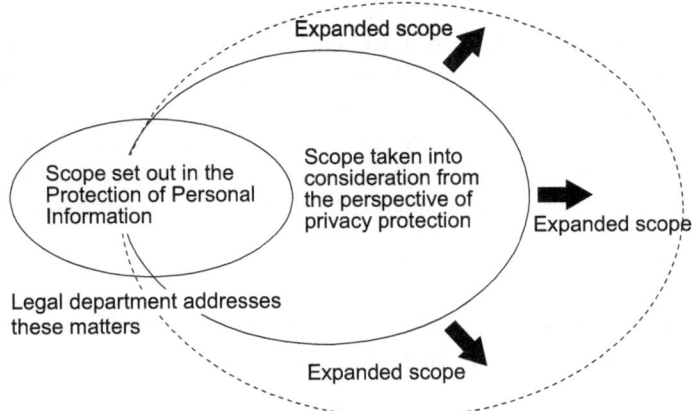

Scope that should be considered from the perspective of privacy protection
This implies that one should go beyond the scope set out in the Protection of Personal Information by applying an extra level of consideration on the assumption that requirements will vary depending on the technology and information used and on socio-environmental circumstances.

Fig. 6.1 Privacy governance

Smart-city services should involve residents from the design stage, ensure that residents are fully informed about the purpose of the service and then brief them about the risks, so that residents will support the service from the piloting phase onward. In some cases, the people whose privacy must be taken into consideration are not the immediate data providers but the observees. In these cases, the community as a whole must be on board with the smart-city service from the design stage forward.

3. *Agile Governance*

Once the service is launched, residents and users may start feeling fresh misgivings. The service may have been designed meticulously and gained the support of residents, but as we have seen, fresh concerns can arise with subsequent technological developments or because of the problems occurring in other cases.

If fresh concerns arise in this way after the service has started using the data or applying AI, a review should be conducted and any issues should be addressed as necessary. Ideally, the service should be designed from the start to be technologically flexible enough to accommodate such changes. If a service uses personal data, for example, it should be designed so that it can stop, after the event, building the data into the service when the data provider requests.

The same goes for services that use corporate data. Because such services often involve a chain of business operators who use the data, the system should be designed so that a business operator further down in the chain will stop using the data if a business operator further up in the chain so requests. If the service is channeled through a data platform, the platform operator will need to contractually compel platform users further down the chain to take measures as necessary to dispel any misgivings parties further up the chain (including observees) may have (Digital Agency & Cabinet Office 2022).

Such measures will entail a change in the approach to governance itself. Smart cities involve a trial-and-error process for building a better community, and data governance involves trial and error too. As the rules being perfect from the start is unlikely, a rapid iterative cycle of applying rules, reviewing how well they are working, and making necessary changes will be required. This type of governance is known as agile governance (Ministry of Economy, Trade and Industry 2022). Agile governance is the type of governance smart cities require for data governance.

6.4 Balance of Benefits and Risks Determines Social Acceptability to Use of Data in Smart Cities

Smart cities depend on the use of data and AI, but as we have seen, the use of data and AI involve risks. Legal or regulatory compliance alone is not enough, and even the best designed system in the world can never be risk-free; residents and other parties concerned must be forewarned about the risks. On this basis, residents and other parties concerned will need to be informed that agile governance in operating the service is being followed, so that they are informed about data use and the approach to governance. Ultimately, whether the residents will accept a service with nonzero risks hinges upon whether they believe that the benefits the data- or AI-driven service will deliver to residents and the community as a whole will outweigh the risks. It is important to avoid putting the cart before the horse. Rather than using data or AI being the goal, the basic idea must be that the data will enable an outstandingly beneficial service and that the benefits of the service can be delivered in no other way.

Local communities face a host of tricky problems. Startups use a business model that identifies and solves a burning need—a need that cannot be ignored, similar to someone's head catching fire. With smart cities, what matters is not using data and AI per se but how effectively the use of data and AI will address the community's severe problems. When the cart is put before the horse (when use of data and AI becomes the end goal), it will not be possible to build social acceptability to data and AI.

The benefits of the service are not always delivered directly to the data providers or observees. The burning needs that a smart city addresses may be less to do with individuals' needs and more to do with serving the common good—addressing the burning needs of society as a whole. Survey data suggests that social acceptability toward providing data can be encouraged in these cases, even though the data providers get no immediate benefit as a quid pro quo for providing their data (Watanabe and Hirai 2022). In other words, as long as the service benefits the community as a whole by addressing the community's problems, and as long as the residents are fully informed of this, then the residents would probably support the service even though the individual members of the community do not benefit directly. Either way, it is important to take a service-first approach—to clarify the services that will be delivered—so that a narrative for tolerating nonzero risk can be created.

References

Act on the Protection of Personal Information (2003) https://elaws.e-gov.go.jp/document?lawi d=415AC0000000057. Accessed March 27, 2024. https://www.japaneselawtranslation.go.jp/ ja/laws/view/4241. Accessed March 27, 2024.

AI Data Consortium (2022) Report on Data Provision Contract for the AIDC Platform. https:// aidata.or.jp/wp-content/uploads/2022/02/Report-on-Data-Provision-Contract-for-the-AIDC-Platform.pdf. Accessed March 27, 2024.

Digital Agency & Cabinet Office Intellectual Property Strategy Promotion Secretariat (2022) *Purattofōmu ni okeru dēta toriatsukai rūru no jissō gaidansu ver 1.0* [Practical Guidance on Data-Handling Rules for Platforms, ver 1.0]. March 4, 2022 https://cio.go.jp/sites/default/ files/uploads/documents/digital/20220304_policies_data_strategy_outline_01.pdf. Accessed March 27, 2024.

Ministry of Economy, Trade and Industry (2021) *Eigyō himitsu: Eigyō himitsu o mamori katsuyōsuru* [Trade secrets: Safeguarding and Using Trade Secrets]. https://www.meti.go.jp/ policy/economy/chizai/chiteki/trade-secret.html. Accessed March 27, 2024.

Ministry of Economy, Trade and Industry, Intellectual Property Policy Office (2021) *Gentei teikyō dēta ni tsuite* [On restricted data]. December 20, 2021. https://www.meti.go.jp/shingikai/ mono_info_service/sports_content/pdf/002_s01_00.pdf. Accessed March 27, 2024.

Ministry of Economy, Trade and Industry, Expert group on how AI principles should be implemented, AI Governance Guidelines WG (2022) Governance guidelines for applying AI principles, ver 1.1. January 28, 2022 https://www.meti.go.jp/press/2021/01/2022012500 1/20220124003.html. Accessed March 26, 2024. https://www.meti.go.jp/shingikai/mono_ info_service/ai_shakai_jisso/pdf/20220128_2.pdf. Accessed March 27, 2024.

Ministry of Economy, Trade and Industry, Study Group on New Governance Models in Society5.0 (2022) Agile Governance Update -How Governments, Businesses and Civil Society Can Create a Better World By Reimagining Governance, Governance innovation Vol. 3. (January 31, 2022) https://www.meti.go.jp/press/2022/08/20220808001/20220808001.html. Accessed March 27, 2024. https://www.meti.go.jp/press/2022/08/20220808001/20220808001-b.pdf. Accessed March 27, 2024.

Ministry of Foreign Affairs of Japan (2023) What is the SDGs? https://www.mofa.go.jp/policy/ oda/sdgs/index.html. Accessed March 27, 2024.

Ministry of Internal Affairs Communications & Ministry of Economy, Trade and Industry (2021) *DX jidai ni okeru kigyō no puraibashī gabanansu gaidobukku ver 1.3* [The Guidebook for Corporate Privacy Governance in the Digital Transformation (DX) Era, ver 1.3]. https://www. meti.go.jp/policy/it_policy/privacy/guidebook_ver1.3_english.pdf https://www.meti.go.jp/ policy/it_policy/privacy/privacy.htmlAccessed March 27, 2024.

Tokyo Shoko Research, LTD., TSR survey (2022) The survey was conducted between January 2012 and December 2021. It aggregated data leakage and data loss incidents in listed companies and their subsidiaries as of the day they were voluntarily reported. https://www.tsr-net. co.jp/data/detail/1191102_1527.html. Accessed March 27, 2024.

Unfair Competition Prevention Act (1993) https://elaws.e-gov.go.jp/document?lawid=40 5AC0000000047. Accessed March 27, 2024. https://www.japaneselawtranslation.go.jp/en/ laws/view/3629. Accessed March 27, 2024.

Watanabe T, & Hirai Y. (2022) *Dēta teikyō ni okeru kenenten to kōeki kōken ishiki kōka no eikyō: COCOA riyō ni kansuru shitsumon hyō chōsa kara* [Contribution in Data Provision by Data Controllability and Awareness of Public Interest: Questionnaire Survey on COVID-19 Contact-Confirming Application]. Journal of Intellectual Property Association of Japan 18(3), pp. 29–38. https://www.ipaj.org/bulletin/backnumber/JIPAJ18-3/p29-38.html. Accessed March 27, 2024.

Chapter 7
Citizen Participation

Tomoyo Sasao and Yuki Igeta

Abstract This chapter examines the critical role of citizen participation in smart-city development, with a focus on the implementation of living labs as a framework for fostering community engagement and co-creation. Living labs are user-centered, open-innovation ecosystems that integrate research and real-life experimentation, allowing citizens to actively contribute to the design and development of urban services. This participatory approach ensures that smart cities remain aligned with enhancing residents' well-being, rather than merely advancing technological solutions. The chapter explores various case studies, including the Kashiwa-no-ha Living Lab, which serves as a model for integrating citizens into the decision-making processes of smart-city initiatives. Key findings highlight the importance of transparency, continuous dialogue, and the democratic involvement of diverse stakeholders. Challenges encountered in managing living labs, such as balancing technological feasibility with community needs and sustaining participant motivation, are also discussed. The chapter concludes by presenting best practices and strategic insights for effectively leveraging citizen participation in smart-city projects. By embracing living labs, smart cities can achieve a more inclusive, sustainable, and people-centric urban environment that reflects the true needs of their communities.

Keywords Citizen participation · Living labs · Co-creation · Smart-city innovation · Community engagement

T. Sasao (✉)
Faculty of Engineering, Reitaku University, Chiba, Japan
e-mail: tsasao@reitaku-u.ac.jp

Y. Igeta
Graduate School of Frontier Sciences, The University of Tokyo, Tokyo, Japan
e-mail: igeta.yuki@edu.k.u-tokyo.ac.jp

© The Author(s) 2025
Hitachi-UTokyo Laboratory (H-UTokyo Lab.), *The Architecture of "Society 5.0"*,
https://doi.org/10.1007/978-981-96-2929-9_7

7.1 Citizen Participation in Smart Cities

In smart cities, data-driven and digitally driven innovations are being developed and rolled out by public, private, and academic organizations. However, how many opportunities have been provided for ordinary citizens to participate in these projects? Many citizens may have no idea of what innovations are underway or may be surprised to find services that they never knew were coming. When a gulf emerges between the smart-city projects and the residents, the projects will start to lose sight of the underlying goal of smart cities—to enhance people's well-being—and end up making the means the end goal. To ensure that smart cities are sustainable and people centric, they must provide opportunities for diverse citizens to participate. Smart-city initiatives must always be transparent to everyone so that residents can understand what is going on; if they involve controversial issues, opportunities should be provided for robust debate to ensure a democratic process for determining the policy for the service in question. The factors that will enhance people's well-being should be identified and incorporated into the next growth strategy, and citizens should proactively involve themselves in the process of developing the services, so that the community's services will be as good as possible.

For smart cities in Japan, the Cabinet Office's 2021 Smart City Guidebook (Cabinet Office et al. 2021) includes citizen participation as a key topic. However, currently, there is no proper approach on how to encourage citizen participation or what form citizen participation should take. In its absence, local communities across Japan have undertaken trial-and-error processes in citizen participation. At H-UTokyo Lab, we have focused on the potential of the living lab, an open-innovation approach that involves citizens and other stakeholders. That is, we have researched the potential of a living-lab framework for encouraging citizen participation in smart cities. This chapter presents the background to living labs and explains why it is good to apply the living-lab approach in smart cities. We have worked as members of the team operating a living lab in cooperation with Kashiwa-no-ha Smart City (in Kashiwa, Chiba), which is known as "Kashiwa-no-ha: a community-development studio for all. (UDCK Town Management 2024)" As members of this team, we have worked since the launch of the living-lab projects. The second part of this chapter presents the work undertaken in this team and clarifies project themes in which living labs can play an effective role in smart cities, and the project-specific best practices for managing a living lab.

7.2 Living Labs: History and Value

The living-lab concept was first introduced in the late 1990s, by William J. Mitchell of MIT (Mulvenna et al. 2011; Eriksson et al. 2006). During that time, the term "living lab" connoted a practice in which researchers observed, analyzed, and evaluated technological applications and their effects in real-life settings (Hossain et al. 2019;

Schuurman et al. 2011). The term also spread from the USA to Europe, leading, in 2000, to the creation of the Botnia Living Lab in Luleå, Sweden (Luleå University of Technology 2024). The year 2006 saw the launch of a large living-lab network in Europe, the European Network of Living Labs (ENoLL) (European Networks of Living Labs 2024). ENoLL defines living labs as "user-centered, open-innovation ecosystems based on a systematic user co-creation approach integrating research and innovation processes in real-life communities and settings." Thus, the initial definition of living labs was broadened with the addition of concepts such as community, co-creation, and innovation. From that point on, living labs engaged in a wider range of co-creation themes, from ordinary objects such as furniture and daily necessities to themes related to community as a whole, such as healthcare and urban planning. The year 2010 saw the first annual conference of Open Living-Lab Days in Paris, a gathering where participants shared their experiences and research in living labs. The number of attendees has increased each year. Thus, living labs are now welcomed around the world and engage in a wide array of practices.

The first living labs emerged in Japan during the 2010s. In 2016, the government emphasized citizen participation in its 5th Science and Technology Basic Plan (Cabinet Office 2016). In 2020, METI released the Living-Lab Guidebook (Ministry of Economy, Trade and Industry 2020). These developments sparked interest in living labs, and today we are seeing a rise in the number of such labs. Living labs are also used for smart cities. Of the 36 communities that MLIT selected for funding under its 2021 program for smart-city model projects and key projects for developing into businesses (Ministry of Land, Infrastructure, Transport and Tourism 2022), three had included a plan to establish a living lab in the smart-city zone: Kashiwa-no-ha Smart City (in Kashiwa, Chiba) (Kashiwa-no-ha Smart City Consortium 2020), Arao Well-Being Smart City (in Arao, Kumamoto) (Arao Smart City Promotion Council 2020), and Smart City Kure (in Kure, Hiroshima) (Kure Smart City Consortium 2021). These examples illustrate how living labs are starting to be used for smart-city programs.

We examined individual examples of living labs in Japan. We noticed some differences in how the living labs were defined; some were designed as *spaces* for co-creation, some as *organizations* managing co-creation, and some as *processes* for co-creation. However, many shared what Akasaka and Watanabe (2022) have highlighted as the key methodological features of living labs: (1) co-creation with users and other stakeholders and (2) experimentation and design practices in real-life settings. These findings suggest that living labs in Japan tend to emphasize the concepts that have been associated with living labs since the turn of the century: co-creation with the community and open innovation.

In view of these findings, we define living labs for smart cities as follows: a series of techniques that ordinary citizens and other stakeholders follow with a common purpose (usually to address a community need); these techniques involve experimenting in real-life settings and co-creating as part of the process of identifying needs, forming a vision, developing ideas and designs, and developing services. Figure 7.1 shows the sequence of activities in a living lab. The process is based on the design thinking process typically seen in various other initiatives. However, it

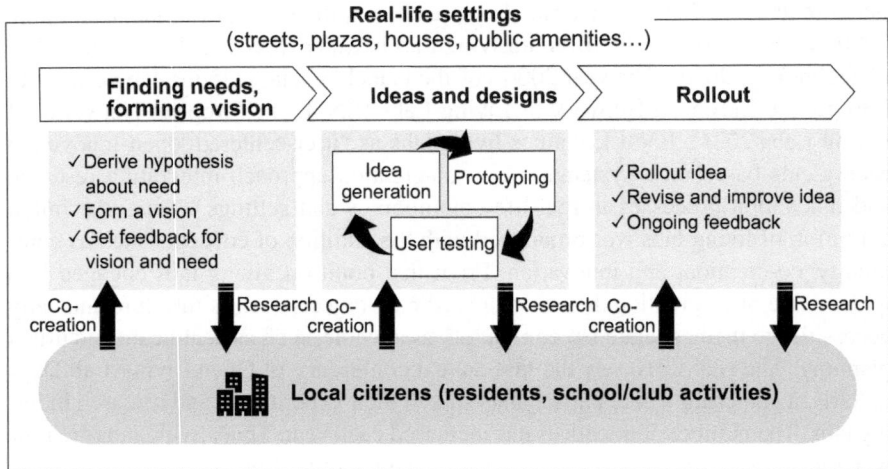

Fig. 7.1 H-UTokyo Lab's conception of the process in living labs for smart cities

differs in two respects: (1) the stakeholders (the company or other organizations managing the services and the ordinary citizens who will use the service in the future) bring their own particular strengths to the development process and (2) a PDCA cycle can be followed in a real-life setting, as the living lab can utilize an advantage of a smart city—the demonstration field.

7.3 Living-Lab Practices: Kashiwa-no-ha Smart City

Kashiwa-no-ha is an area in the Tokyo suburbs. It is situated in the northwest of Kashiwa City in Chiba Prefecture, 30 kilometers from central Tokyo. In 2005, the Tsukuba Express rail link was opened, enabling access to or from central Tokyo in 30 min. In tandem with the construction of the rail link, redevelopment projects were launched in the areas served by the rail line. In the Kashiwa-no-ha area, a number of amenities were established to create an urban hub; these included a campus of the University of Tokyo, a campus of Chiba University, and national research institutes. The creation of these amenities set the stage for the launch, in 2019, of the Kashiwa-no-Ha Smart City Consortium, consisting of 19 organizations, which included Kashiwa City, Mitsui Fudosan, and UDCK. This consortium has led smart-city initiatives related to mobility, energy, public spaces, and wellness in the zone covering a 2-kilometer radius from Kashiwa-no-ha Campus Station (Kashiwa-no-ha Smart City Consortium 2020).

Kashiwa-no-ha Smart-City's action plan defines a living lab as a new kind of process, one that involves citizen participation and open innovation. In December 2020, it launched its own living lab, naming it "Kashiwa-no-ha: a community-development studio for all" (we shall refer to it as the Kashiwa-no-ha Living Lab) (UDCK Town Management 2024). The living lab has developed a number of programs with the goal of having ordinary citizens (the users of smart-city services) take the lead and co-create Kashiwa-no-ha Smart City with government, the private sector, and academia (see Fig. 7.2). The Kashiwa-no-ha Living Lab is managed by UDCK (which also heads the Kashiwa-no-ha Smart City Consortium), UDCK Town Management, and—bringing an academic perspective—H-UTokyo Lab. Operating as the secretariat of the living lab, this management team is always looking for a better methodology.

Introduced below are four programs the living lab conducted between 2020, when the lab was first launched, and 2022. These examples illustrate the diverse roles smart cities can play in smart cities. They also illustrate the problems the secretariat has experienced and matters it has learned in managing the living lab, which we hope will serve as a guide to what you need to be aware of when launching a living lab.

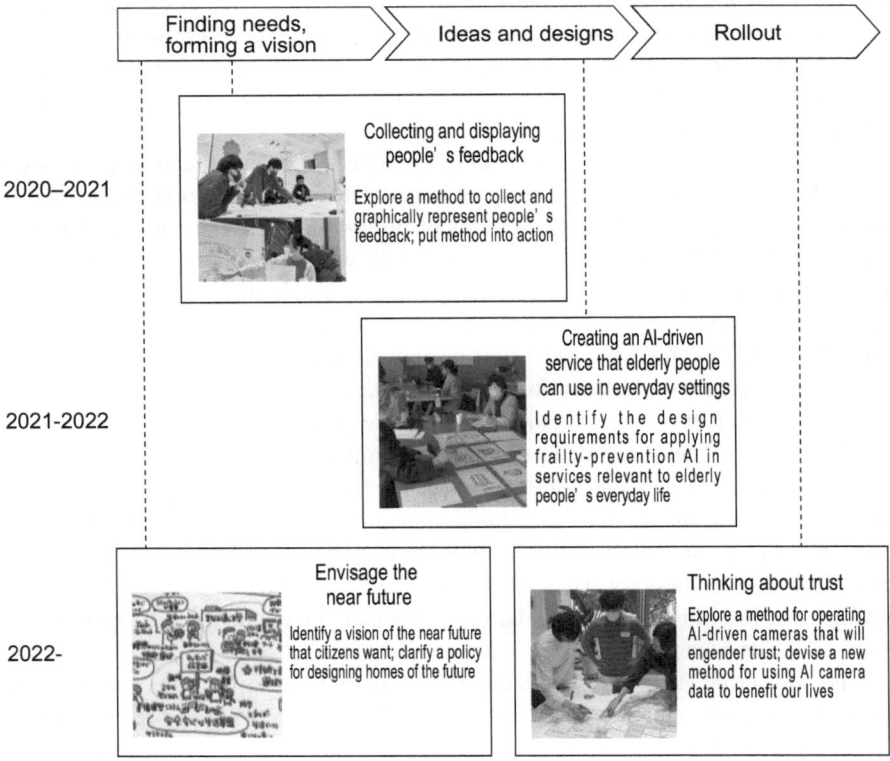

Fig. 7.2 Four living-lab programs at Kashiwa-no-ha Living Lab

7.3.1   Program 1: Collecting and Displaying People's Feedback

Kashiwa-no-ha Living Lab's first program was named "collecting and displaying people's feedback." This program was launched amid concerns of a lack of contact and communications with citizens, although UDCK had, since 2006, tried to reach out to the Kashiwa-no-ha community and continued efforts to involve diverse citizens into the program. The reason why we set this specific theme is that there was a strong desire for a more people-centric approach (one that was sensitive to what residents were thinking and how they wanted the community development to proceed) and mounting concerns that the residents had remained aloof from the smart-city projects. In this maiden program, we would work with citizen participants to devise, and then implement, a system for collecting and displaying resident feedback. Eventually, the system would be used to derive from the feedback the hidden needs of residents and residents' hopes for the future. With these insights, a more grass-roots approach could be taken when engaging with Kashiwa-no-ha Living Lab and Kashiwa-no-ha Smart-City's project themes.

In the first 6 months, the lab recruited members of the public who were interested in the project. It recruited 19 participants from Kashiwa City and elsewhere. The lab first ran some activities to build rapport with the participants and familiarize them with the process of dialog. The participants then presented their thoughts to the community. Four separate ideals were identified: ambition, discovery, learning together, and making friendships.

For each of these ideals, participants separated into teams and brainstormed ideas. The participants got a real sense of both the difficulty and good results of collecting extensive feedback in the process of prototyping ideas and testing them in real-life settings (such as prototyping a simple website, collecting feedback on social media, and organizing community dialog events). In their final presentations, the five teams revealed their ideas, along with the results of prototyping and testing. We then invited people from local startups, government bodies, and research organizations to discuss the outlook for developing the participants' ideas.

Some of the teams proposed ideas for activities that the participants themselves would maintain. These included a system of communication in which people would contribute entries to a group journal on social media and the provision of opportunities for instant events that would obtain spontaneous feedback from people who drop in to the event. Some of the teams proposed ideas that will only become feasible with new technology. These included installing digital signage in the station vicinity and providing an online platform on which anyone could post comments, vote, or debate ideas.

The secretariat (in charge of planning and running the living lab) was worried about how the participants' ideas could be applied in practice. We had hoped that the discussions following the final presentations would be attended by people from companies that could put the ideas into practice. However, although the ideas looked good on paper, their feasibility as business propositions were another matter. Consequently, the ideas that were designed to be run by the participants themselves were continued, with the participants starting to put the ideas into practice. However, the ideas that were designed to be run by the living lab as a whole were put on hold in the absence of any clear vision for the next step. After a year had passed, one of the latter ideas gained backing from an entirely different source. At a startup conference in Kashiwa-no-ha, a startup proposed a prototype that matched the proposed concept of collecting and displaying people's feedback, raising hopes that the participants' idea could be put into practice, even if only in part. After we presented Kashiwa-no-ha Living Lab's proposed idea, the startup tweaked its prototype and agreed to pilot a version that was localized to Kashiwa-no-ha. The startup has now reached out to the participants who had thought up the initial idea, and together they have customized the startup's prototype to fit the program theme (collecting and displaying people's feedback). The piloting remains underway.

We learned some lessons from our experience in this program. We learned that any idea pitched in a program may come to nothing unless there are prospects from the start that the idea could get seed funding. Seed funding often proves elusive when you start projects from the needs of ordinary citizens. The lack of tangible output can cause significant stress for participants who engage in a program for years, hoping their efforts will benefit the community. Therefore, when organizing a program that begins from the needs of ordinary citizens, you must fully inform participants about the risk that their ideas may never take shape at the early stage and that the program could drag on for a long time.

I do not want to give the impression that programs that begin from people's needs are prone to going awry. As we learned in this case, after some time passed, the idea ultimately gained a backer. Without the living-lab's program, this idea would never have intersected with a startup's pilot initiative. Likewise, without giving citizens the opportunity to express their wishes, the startup would not have adapted its system to community needs. Therefore, the organizers of a living lab must never lose hope and keep the antennas up—to keep searching for a good match. A program is more likely to advance to the next stage if the community affords plenty of opportunities for serendipitous encounters with startups and other seed capital.

7.3.2 Program 2: Creating an AI-Driven Service That Elderly People Can Use in Everyday Settings

H-UTokyo Lab has developed an AI application for preventing frailty (more on this in Chap. 12). The app's AI model analyzes big data on healthcare and nursing needs (from the national insurance database and other sources) to predict the person's risk of becoming frail. The app then refers the user to services that can deliver frailty-prevention therapy. The app's main audience was to be elderly people, and we wanted the app to be designed in such a way that elderly people could use the digitally powered services in their everyday settings on an ongoing basis. To that end, we needed to reach out to elderly people and obtain their feedback about the app's usability. In the case of an AI-powered frailty-prevention app, it is relatively straight-forward to work out how people feel about sharing their health data, when it is least stressful for people to enter information about their current status, what input method they prefer, and what kind of output will encourage users to engage in practices for preventing frailty. We launched the renewed program to answer these questions and identify the design requirements that would ensure that elderly users can use the app effortlessly in everyday settings.

In the first program, we had used social media and pamphlets to invite general members of the public to participate. In this second program, we used a different recruitment method. Because the app users would be limited to people aged 70 or older, the participants in this case were recruited by A-Shi-Ta Community Health Promotion Laboratory, a center for participatory health promotion that is located in Kashiwa-no-ha and that has more than three thousand registered members. Because we already had some idea about how the service would be run and who would be involved in supporting the therapy, the center recruited the family members of elderly people, healthcare professionals (doctors, pharmacists, physiotherapists, among others), people who deliver home-visit nursing services or meal delivery services (or who interact regularly with elderly people in their work), and people who have conducted such work or have experienced such roles in the past. The center also recruited public-sector workers as observers. The program included three sessions, each with nine to twelve participants representing diverse perspectives. The participants highlighted the problems and difficulties elderly people experience in everyday life and offered frank feedback about the ideas of the app and about their experiences with using the app prototype. The app development team, which consisted of engineers and members of the secretariat, incorporated this feedback little by little in the design of the service. Eventually, three ideas emerged, each assuming a separate scenario in which the app is used. The participants evaluated the ideas from their various perspectives. As a result, we had identified the requirements we needed to satisfy to make the service viable by the time we completed the program.

During this program, the secretariat encountered new quandaries. The first quandary related to the fact that the first program had started from zero (from citizens' needs) and the second program had begun with the technology already developed to some extent. The first program had allowed for blue-sky thinking; the participants could exercise their creativity and bounce ideas off each other. Such a format is enjoyable and also gives larger independence, and thus participants are more likely to be motivated to continue participating. By contrast, the second program focused on one-way feedback, meaning that participants felt less motivated to continue or to see the point in participating.

The second quandary was a discrepancy between our desire to generate a hypothesis and the feasible program under the constraints. One of the things we had wanted to clarify from the outset was how elderly people would feel about sharing their personal data as a quid pro quo for accessing the service, but we had also predicted that we would get no useful feedback from a handful of elderly people, who were usually scarcely conscious of data matters. Therefore, we decided that with regard to data sharing, we would instead rely on the responses of 1500 elderly people whom we had surveyed in advance. In another example, engineers in the app development team had requested a quantitative evaluation of whether people would continue to undertake the recommended practices for preventing frailty based on the idea of the service they examined. This evaluation would require an intervention study in which participants use a prototype app (when the app is close to becoming a finished product) in everyday settings. However, we had to abandon plans for the intervention because the prototype was not technologically mature enough, and the team lacked expertise in designing an intervention study and getting it approved by the medical ethics committee. In summary, although we had numerous questions about service development, constraints at the time allowed us to address only a limited subset. This experience highlights the importance of identifying and prioritizing key questions at the outset of program design.

7.3.3 Program 3: Envisage the Near Future

The first two programs were supported by a clearer idea about what the service in question should be. The program we worked with the National Institute of Advanced Industrial Science and Technology was different from the above programs. The purpose of this program concerned R&D into digital technologies and living environments that support aging in place (the ability to continue living independently in one's home and community regardless of age). In such R&D, a gap tends to emerge between the design team's vision and the actual needs of citizens. Thus, we wanted to establish a design policy for R&D that would close this gap. We would do so by having people participate in the living lab and clarifying what they want the living environment to be in the near future. This program was designed not to generate an idea for a particular service. It was a new type of program, one that would clarify future living environments that would offer elderly people hope and value, thereby allowing them to age in place.

We recruited 17 elderly people representing different ages, sexes, and family composition. The previous programs had concerned people's present needs, whereas this program concerned the near future, which required participants to envisage a life that they had never experienced in reality. Therefore, we designed the program to account for this. That is, we organized a sequence of sessions that would help participants gain an increasingly specific idea of what their living environment would need to be for them to maintain their mental and physical well-being. For example, in one session the participants split into pairs and took turns to interview each other about what the living environment might be 10 years from now. In another session, participants drew images depicting the future relationships they will have at age 75. In another session, an expert delivered a lecture about the challenges people encounter as they age. By the end of the program, the participants had identified four future scenarios that aging-in-place services should adopt: positive aging, semi-loose connections, circle of friends, and quality solo life.

Initially, we were concerned that participants might struggle to envision realistic future scenarios, as they had never experienced such a future. However, the living lab characteristics—particularly the meticulous, iterative process for designing a program—proved well suited to the participants' task of envisaging future scenarios. Citizens cannot plunge into the task of envisaging the future by themselves. Participants were steadily primed to envisage a realistic scenario by having experts brief them on the physical and cognitive changes that occur with aging, by presenting them with evidence about living environments and future technologies, and by identifying the values of diverse participants. The participants' realistic future visions, refined by the ideas of experts and other participants, testify to the effectiveness of co-creation and also represent one kind of output that living labs are capable of.

7.3.4 Program 4: Thinking About Trust

The final program to discuss is one that focused on a service that had already been launched. The program involved 29 AI-driven cameras that were installed in the vicinity of Kashiwa-no-ha Campus Station in 2021. The cameras are designed to be sensitive to people's privacy: rather than uploading all footage to a server, the cameras recognize when an abnormality is occurring (such as someone falling, crouching down, or brandishing a weapon) and estimate the age and gender of the person. When rolling out technology that requires privacy considerations, we need to ensure that the system is running properly, but also that its actual state is transparent and that the public are aware of this and can trust it. Before installing the cameras, Kashiwa-no-ha organized briefing meetings and distributed leaflets to inform the public about them, but it had no opportunities to find out how residents perceive the design of the cameras. Against this backdrop, the fourth program was designed to enhance residents' understanding of the role and functions of the AI cameras, to think together about the requisites for trustworthy operation, and to consider how the cameras' functionality could be expanded to enable new applications that can contribute to residents' well-being.

When engaging in the task of examining a trustworthy operation method, we collaborated with the World Economic Forum's Center for the Fourth Industrial Revolution Japan (C4IR Japan), which is interested in citizen-led trust governance. C4IR Japan assisted us by delivering a citizen workshop using a risk chain model developed by the University of Tokyo's Institute for Future Initiatives (Institute for Future Initiatives, The University of Tokyo 2021). The risk chain model was a workshop tool that was originally designed for an AI application's development and operation teams to use in cross-disciplinary workshops to identify the risks and steps to counter risks associated with the service. In this case, however, the model was adapted so that it could be used by lay citizens. The participants used the tool to analyze the risks that exist when the AI cameras are operating and the steps to counter these risks. Participant feedback revealed that the workshop helped the citizens understand the technological and operational matters. In a questionnaire, the participants reported that the tool helped them clarify the roles and functions of the cameras and to understand the roles of and relationships between the stakeholders (data scientists, the secretariat, and users); 73% of the 15 participants said that the risk chain model proved useful in helping them thinking about AI camera operation.

The next task was to consider how the AI camera's functionality could be expanded to explore new methods of utilization for the sake of their own well-being. The participants formed into teams of members who share similar ideas about the community's needs and a similar future vision for the community. For around 6 months, the teams brainstormed and prototyped ideas. One team proposed an app with useful information for dog owners. Dog walkers could use the app to see street temperatures, the level of congestion in amenities that allow pets, and the locations of other dog walkers. Another idea involved measuring the brightness of people's moods and displaying the results on digital signage outside the station to raise people's awareness about the AI cameras. One other app would involve making a hybrid of AI camera lenses and residents' eyes and having the city collaborate with redevelopment companies in carrying out repair work in the community. The final presentations were attended by eight guests, including an AI camera business, a city worker, and an expert in community design. These guests gained insights from the presentations, and the findings are promising for the future activities; in a questionnaire, 75% of the guests answered that they could relate the presentations to their work and that they would prefer to back the ideas.

Figure 7.3 presents the results of pre- and post-program surveys (both with identical questions), revealing how the participants' attitudes to AI cameras changed between the two time points. The responses suggest that the program gave participants a better understanding of the usefulness of AI cameras, eased their concerns, and enhanced their trust. Notably, we observed a 29-point increase between the two surveys in the percentage of participants who believed that AI cameras are useful to others, suggesting that the program, by offering encounters with a range of different people's values, fostered a sympathetic attitude toward other participants. Anxiety about AI cameras was low; even in the first survey, no more than 20% of participants expressed such anxiety. Trust in the management team/operator was high (exceeding 70%). The program neither eased anxieties nor improved trust in some participants. In such cases, the causes of anxiety must be identified and alternative care approaches considered.

7.4 Design Points for Running Living Labs in Smart Cities

Figure 7.4 presents the themes to which living labs are suited according to the findings of the above case studies and the best practices in Kashiwa-no-ha. The themes are divided into three broad categories: service co-creation, future vision, and fostering understanding.

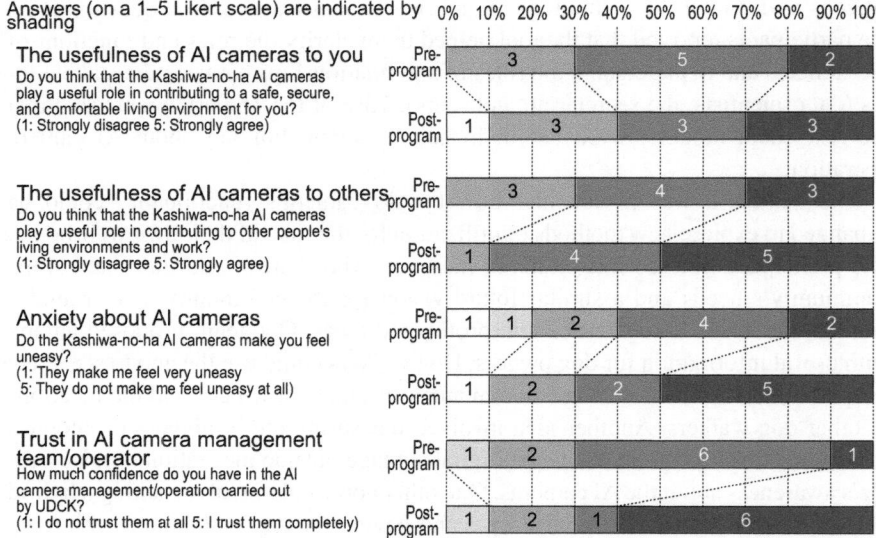

Answers (on a 1–5 Likert scale) are indicated by shading

Numbers in the bar graph indicate number of responses.

Fig. 7.3 Results of surveys on participants' attitudes to AI cameras (the first survey conducted before and the second after Kashiwa-no-ha Living Lab's program)

Theme category	Theme	Living-lab objective	Example from Kashiwa-no-Ha Living Lab
Service co-creation	Co-creating a service related to ordinary citizens' needs and ideas	- Exploratory approach e.g. How can Citizens' well-being be improved?	1,3
	Co-creating a service driven by technology developed by a company or smart city	- Confirmatory approach e.g. Does this function improve citizens' well-being?	2,4
	Co-creating a service intended for a certain demographic	- Improve usability - Explore touchpoints with real worldex. e.g., Where, and how should it be used?	2
	Co-creating a service related to a complex socio-environmental problem requiring multi-stakeholder collaboration	- Social innovation - Open Innovation e.g. Who will collaborate with whom and how?	1,2,4
Future vision	Envisaging a future community and living space	- Perceive the advancing technologies and services that will be required in the future e.g. What should be developed in the future?	3
Fostering understanding	Knowing and understanding the future service structure, technologies, and datasets that will be used	-Raise public awareness about a smart-city initiative -Foster assurance and trust e.g. Do you know about your community's smart city?	4

Fig. 7.4 Themes that have high potential for living labs in smart cities

The majority of the themes fall under the service co-creation category. Figure 7.4 includes four examples of services for this theme. As the objectives will differ somewhat in each theme, it is important to tailor the design of the program to the attributes of the theme in question. The Kashiwa-no-ha programs corresponding to this theme include the program to collect and display people's feedback and the program to develop an AI-driven frailty-prevention service.

The next theme is future vision. Rather than involving services that are already in operation, programs under this theme involve envisaging the future of the community or people's future living environments and then anticipating their desire to contribute to the future technological development. Unlike in the case of service co-creation, the programs must be designed in a way that participants can easily envisage the future scenario, so bear this point in mind when designing the program. Of the Kashiwa-no-ha programs, the program for envisaging a living environment in the near future corresponds to this theme.

The third theme is fostering understanding. The purpose in this theme is to improve the public's understanding about a smart-city service in operation. It also involves programs that are designed to assure the public and foster trust. In smart cities, such programs will become increasingly important and the needs will increase. Kashiwa-no-ha's program on AI-driven cameras corresponds to this theme.

We identified these themes based on past practices, but as living-lab projects proliferate in smart cities across Japan, we might see additional themes emerge. Given this, living labs should keep exchanging opinions with people practicing on sites and keep updating the set of themes that are effective for living labs in smart cities to engage in. The next section introduces the problems that are likely to occur when running projects in each theme (with reference to the four different types of services under the co-creation category of themes) and discusses solutions and design points for addressing these problems, as gleaned from our own practices. Hopefully you will find this discussion a useful guide in your own living-lab practices.

7.4.1   Service Co-creation

1. Co-creating Services Related to Ordinary Citizens' Needs and Ideas

With many smart cities today having the objective of increasing well-being, smart-city initiatives and services are increasingly being designed from zero—in a process that begins with the ideas and needs of ordinary citizens. For these smart-city projects, it makes sense to design living-lab programs that take an exploratory approach and glean insights from ordinary citizens.

However, as demonstrated in the "collecting and displaying people's feedback" program, exploratory approaches often face a challenge: the needs of citizens do not always align with the emerging technological innovations intended to serve them. For such programs, participants must understand the risk that their ideas may never take shape and the possibility that the program will drag on before starting their cooperation in the program. We may also have to bolster the organization running the living lab; this organization should have a matching function and channels with a wide range of startups and other business operators, so it can create plenty of opportunities for program outputs (ideas) to be taken up by companies that can implement them.

Another common problem in such programs is that the participants sometimes fail to examine the business model (a model for ensuring profitability) and technological feasibility. Ordinary citizens are experts when it comes to their own lives, but they often lack expertise in how to monetize services and the technology itself. One way of addressing this problem is to bring on board an interested company or research institute once an idea has been generated. The participants can then discuss the idea together with the organization so that the idea can be developed into more feasible one.

2. *Co-creating a Service Driven by Technology Developed by a Company or Smart City*

The vast majority of smart-city projects today consider developing a service that uses existing core technology. For these cases, there are three main things to be aware of.

The first thing is that living labs engaging in such projects are prone to treating the participants as mere technology monitors. When all participants are doing is providing one-way feedback, they are unlikely to feel that they are participating in co-creation and their motivation tends to decrease as a result. Participants need to understand that the project will deliver substantial changes to their lives and feel that they have a stake in the project. For this reason, at the start of the program, participants must be informed about the purpose of co-creation and the value of having ordinary citizens' participation.

The second thing to be aware of is the difficulty of setting the parameters for idea generation. If the technology underlying the idea generation is specified too precisely, then only the ideas within the constraints of the technology's present functions will be obtained. However, if participants have a poor grasp of the technology in question, the ideas will be too unrealistic to develop into a tangible output. To maximize the effectiveness of co-creating a service for which a technology to utilize has been decided on with ordinary citizens, we need to think carefully about the parameters and scope for the idea generation—the ground rules for thinking up ideas and the extent to which such idea generation can be included in hypothetical thinking. The program should also pose some questions to stimulate and guide the participants in generating ideas. These questions might include "what additional features should be added?" or "what technological bottlenecks might we face in expanding the scope of applications?"

The third thing to be aware of when co-creating a service driven by the technology developed by a company or smart city is that such co-creation involves vast volumes of personal and sensitive data, and how the data should be handled is never clear based only on the opinions of the ordinary citizens participating in the living lab. The government has produced guidelines on this matter, but in many cases the legal and moral implications for the case in question are unclear. One must take measures to address this ambiguity as early as possible. Such measures may include creating structures for expert advice (such as providing a dedicated ethics committee and creating a system in cooperation with legal experts) and applying to have the area designated as a special strategic zone (which would enable innovative initiatives in the area). The living lab cannot do everything by itself, so the smart-city's resources will need to be leveraged.

3. *Co-creating a Service Intended for a Certain Demographic*

In smart cities, many of the initiatives and services are for the general public as a whole, but some are for a more specific demographic such as parents or the elderly. To ensure that the service is accessible to the target audience, the living-lab's participants should be from that audience. The problem here is that the target demographic might be hard to reach by the usual channels of social media and leafleting. In the case of the frailty-prevention app developed in program 2 (creating an AI-driven service that elderly people can use in everyday settings), the target demographic was people aged 70 or older. For that reason, we recruited participants from a Kashiwa City community center (A-Shi-Ta Community Health Promotion Laboratory) whose membership included many elderly people. As this example illustrates, when it comes to recruiting participants, the living lab needs to have connections with communities containing the target demographic.

Even if we are not fully sure what the intended audience of the service will be at the outset of the program, the target demographic may be narrowed in the course of discussing ideas. This was the case, for example, in program 4 (thinking about trust). After the program got underway, participants pitched an idea to add a function to the AI cameras to deliver a service for dog owners. To confirm whether there was such a need, we conducted a survey among local dog owners. In retrospect, if we had a connection with a local dog owners' club, we might have been able to include a more diverse array of perspectives and better co-creation. When the organization running the living lab has connections with different communities (elderly people, parents, junior-high and high school students, dual-income couples, and so on), then when participants suddenly generate an idea for a particular demographic, members of that demographic can be easily invited to participate in the process, resulting in more effective co-creation.

4. *Co-creating a Service Related to a Complex Socio-Environmental Problem Requiring Multi-Stakeholder Collaboration*

Multi-stakeholder collaboration is required to address complex socio-environmental problems such as the spiraling costs of healthcare in an aging society, the need to maintain infrastructure in provincial cities with dwindling populations, and the need to manage disaster risk in hazard areas highlighted in hazard maps. Co-creation is more effective when the different stakeholders contribute their particular strengths to the process. To that end, a wide array of stakeholders organizationally embedded in their roles needs to be considered. Figure 7.5 illustrates an

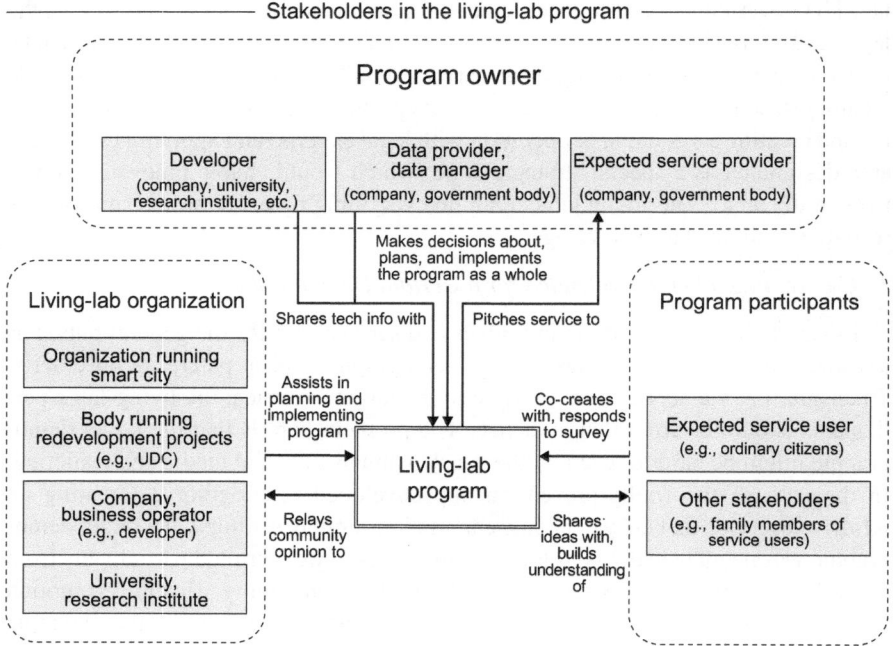

Fig. 7.5 Illustration of ideal inter-stakeholder relationships in a living-lab program for creating an AI-driven service that elderly people can use in everyday settings

ideal network of inter-stakeholder relationships as gleaned from our experience in program 2 (creating an AI-driven service that elderly people can use in everyday settings). First, there are three broad roles in a living lab. The first is the organization responsible for implementing the living-lab program. This party introduces the theme of interest and provides participants with the information and materials (the underlying technology, for example). The second is the living-lab organization. This is the organization that runs the living lab as a whole and supports the organization responsible for implementing programs by managing, facilitating, and advertising the living lab's programs. Third is the program participants. These include the service users (and their family members) and other citizens who may have a stake in the theme. The figure includes the stakeholders who fulfill each of these three broad roles, but the stakeholders may vary depending on the stage of the program, so such a structure is not needed at the start of the program. That said, the more complex the theme becomes, the more the stakeholders must be aware of the role they play in the program.

A careful design to fit the demographic traits of the identified stakeholders is also needed so that they can engage in the program. This will require some variation and customization. Working parents and students, for example, may have little opportunity to attend a workshop during daytime hours. For these cases, it might be better to solicit their opinions by an interview or online survey.

7.4.2 Themes in the Future Vision Category

Some smart-city projects are designed to be implemented immediately, whereas others envisage technologies and services that will match the community some decades later. For the latter type of projects, the living-lab program needs to have a long temporal horizon in which the vision or idea is as comprehensive as current technologies allow and paused until the necessary technology becomes available.

As mentioned in the discussion of program 3 (envisage the near future), an issue likely to occur when envisaging a future service is that the participants must imagine what their lives will be some decades from now. Because the task involves thinking about scenarios of which they have no actual experience, participants may experience stress in having to state their opinions or generate ideas. To help the participants envisage a future scenario as specifically as possible, it is particularly important that they are presented with some data about the predictable changes in their physical condition and in the environment and society. It is also helpful to have participants of different ages and gender talk to each other about their daily lives.

A common issue with long-running programs in living labs is that it is hard for the participants to maintain their motivation for co-creation. When part of a program ends in a short space of time and there is a long gap before the output reaches the implementation phase, participants, unable to see their inputs being put to use, may feel that their participation was a waste of time. For this reason, if a company takes with them an idea generated in a living lab with a view to eventually implementing the idea, the participants should be updated on the progress one by one so they have an idea of where their idea is heading. An idea generated in a living-lab program does not have to be turned into one practical application only; it is better to make the idea open knowledge (by broadly publicizing it) so that many organizations can put the knowledge into use. This approach will help in ensuring that the living lab creates benefits for the wider community and society.

7.4.3 Themes in the Fostering Understanding Category

Program 4 (thinking about trust) exemplifies an important type of theme for living labs: building awareness and understanding about a smart-city initiative or service that uses innovative technology or data. By fostering understanding, the living lab can enable people to understand the usefulness of the service, alleviate fear, and build trust in the organization running it. However, when explaining how a service works, making laypeople understand the technology in detail is almost impossible. For this reason, rather than focusing on the technical details, living labs should focus first on getting people to understand the general concept of the service and the situations in which it is used. Although making people understand invisible data or technical minutiae may be infeasible, communicating the core matters of the service, such as those that relate to users' privacy, without dumbing down is important. Asking a science communicator to help in preparing effective transmission designs for a general audience is a good idea.

When engaging in fostering understanding themes, living labs can design the program so that the program, in addition to promoting awareness and understanding, fosters assurance and trust in the participants. Program 4 (thinking about trust) included group work in which the participants identified the risks associated with AI-driven cameras, revealing sources of unease about the data and technology. In this way, the program helped reduce misgivings about the technology and build trust in the operator. Thus, in addition to focusing on informing people about the service and technology, the program should incorporate, as part of this purpose, opportunities for ordinary citizens to express their fears and concerns.

7.4.4 Putting the Living Lab into Action

That concludes the discussion of the three types of living-lab themes, including the issues associated with them and ways to address these issues. Hopefully the discussion has offered insights that can be applied in your living-lab practices.

The final part of this chapter presents an interview question sheet that H-UTokyo Lab composed (Fig. 7.6). These questions are designed for interviews that the living-lab organization will conduct with the program owner during the design of a living-lab program at the initial phase of the program, which is particularly important when the living lab is gaining momentum. This tool can help facilitate co-creation with the program owner in the living-lab program. To design an effective program, any potential areas of conflict or disagreement need to be identified in advance and matters clarified in writing. Programs can involve many stakeholders, but when a common understanding is reached in advance, the program will nonetheless stay on track and the different stakeholders will know each other's roles and communicate with each other smoothly. Outlined below is the interview procedure.

1. Ask about what they ultimately want to accomplish through proposed program themes and write the response in *(1)* in Fig. 7.6. The expected effects in the community and living environment must be described from the perspective of ordinary citizens, as opposed to a company-centric perspective.
2. In *(2)* *(in the figure)*, write down the goal of the program. In *(3)*, write down all that needs to be confirmed during the program. Remember that the ultimate goal written in *(1)* will seldom be the same as the goal stated in *(2)*. A living-lab program has limits; a goal and measurable deliverables that are feasible within the constraints of time, budget, and personnel are needed.

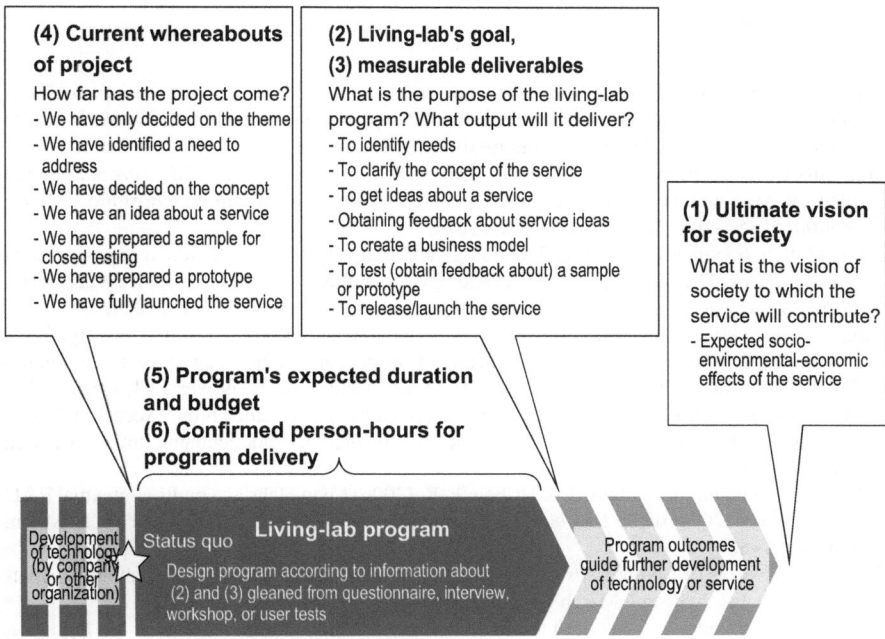

Fig. 7.6 H-UTokyo Lab's interview sheet for initial phase in living-lab programs

3. If the proposed program concerns a service that is underdeveloped or already in operation, ask the owner about how far the service development is in progress, which are ready and to what extent, which are not yet ready, and when the service will enter operation. Write down this information in *(4)*. This information will help to identify the starting point for the living program.
4. The program will involve closing the gap between the living-lab's goal and the status quo. When designing the program, think about how long it will take, what the budget should be, and what the workload will be. Write down this information in *(5) and (6)*.

This chapter presented the living-lab programs in Kashiwa-no-ha to illustrate an approach to encouraging citizen participation in smart cities. I then discussed the common themes for living-lab programs in smart cities along with the issues associated with these themes, ways to address these, and other design points. When it comes to building a smart city together with ordinary citizens, we recommend starting a living lab and using this book as a guide. If one is thinking of launching an organization that will run the sustainable living lab in a community, a system of this organization will need to be built. Descriptions about six key factors for digital society infrastructure can be found in Chap. 14.

References

Akasaka F. & Watanabe K. (2022) *Ribingu rabo jissen no tame no infurasutorakucha kōchiku ni mukete* [Toward building infrastructure for living lab practices]. Journal of the Human Interface Society 24(1): 1–14. DOI: https://doi.org/10.11184/his.24.1_1

Arao Smart City Promotion Council (2020) *Arao werubī-ingu sumātoshiti jikkō keikaku* [Action plan for Arao well-being smart city]. October 2020. https://www.mlit.go.jp/toshi/tosiko/content/001514524.pdf. Accessed March 28, 2024.

Cabinet Office (Council for Science, Technology and Innovation) (2016) *Dai 5 ki kagaku gijutsu kihon keikaku* [The 5th Science and Technology Basic Plan] (Released on January 22, 2016). https://www8.cao.go.jp/cstp/kihonkeikaku/index5.html. Accessed March 28, 2024.

Cabinet Office, Ministry of Internal Affairs and Communications, Ministry of Economy, Trade and Industry, Ministry of Land, Infrastructure, Transport and Tourism and Smart City Public-Private Partnership Platform Secretariat (2021) Smart city guidebook, ver 1.01, Chapter 2-1, April, 2021. https://www8.cao.go.jp/cstp/society5_0/smartcity/index.html. Accessed March 28, 2024. https://www8.cao.go.jp/cstp/society5_0/smartcity/olddocument.html. Accessed March 28, 2024.

Eriksson, M., Niitamo, V., Kulkki, S. & Hribeinik, K. (2006) Living labs as a multi-contextual R&D methodology, 2006 IEEE International Technology Management Conference (ICE),Milan, Italy, pp. 1–8.

European Networks of Living Labs (2024) What are living labs? https://enoll.org/about-us/what-are-living-labs/. Accessed March 28, 2024.

Hossain, M., Leminen, S. & Westerlund, M. A. (2019) Systematic review of living lab literature, Journal of Cleaner Production, Vol. 213, March 2019, pp. 976–988.

Institute for Future Initiatives, The University of Tokyo (Technology governance research unit, AI governance project) (2021) Risk Chain Model (RCModel) Guide, ver 1.0., June 2021, https://ifi.u-tokyo.ac.jp/en/wp-content/uploads/2021/07/RCM_guide_E210706.pdf. Accessed March 28, 2024.

Kashiwa-no-ha Smart City Consortium (2020) *Kashiwa-no-ha sumātoshiti jikkō keikaku* [Kashiwa-no-ha Smart City action plan]. March 2020. https://www.kashiwanoha-smartcity.com/fdownload.php?ctr=download&d=1. Accessed March 28, 2024.

Kure Smart City Consortium (2021) *Kure sumātoshiti moderu jigyō jikkō keikaku* [Actionplan for Kure smart city model project]. https://www.city.kure.lg.jp/uploaded/attachment/59543.pdf. Accessed March 28, 2024.

Luleå University of Technology (2024) Botnia Living Lab https://www.ltu.se/en/research/research-subjects/information-systems/botnia-living-lab. Accessed September 4, 2024.

Ministry of Economy, Trade and Industry (2020) *Ribingu rabo dōnyū gaidobukku* [Guidebook for creating a living lab] March 2020. https://www.meti.go.jp/policy/servicepolicy/living_lab_tebiki_a4.pdf. Accessed March 28, 2024.

Ministry of Land, Infrastructure, Transport and Tourism (2022) *Senshin chiiki no sumātoshiti jikkō keikaku torikumi naiyō* [Action plan/action details for smart cities in model communities] https://www.mlit.go.jp/toshi/tosiko/toshi_tosiko_tk_000051.html. Accessed March 28, 2024.

Mulvenna, M., Martin, S., et al. (2011) TRAIL Living Labs Survey 2011: A survey of the ENoLL living labs, Ulster University.

Schuurman, D., De Moor, K., De Marez, L. & Evens, T. (2011) A living lab research approach for mobile TV, Telematics and Informatics, Vol. 28 no. 4, November 2011, pp. 271–282.

UDCK Town Management (2024) *Minna no machizukuri sutajio* [A community-development studio for all]. https://www.udcktm.or.jp/studio/. Accessed March 28, 2024.

Chapter 8
Smart City QoL-Based Assessment

Kei Suzuki, Mitsuharu Tai, and Tomoyo Sasao

Abstract This chapter highlights the significance of assessing smart cities from a citizen's perspective to foster a society where residents experience happiness. It is crucial to measure quality of life (QoL) and satisfaction with the everyday activities of residents for this kind of assessment. The proposed metrics, ActiveQoL, is designed to measure subjective satisfaction in daily activities including sleep, work, meals, and leisure. ActiveQoL considers four key factors: a fondness for the activity, time spent on the activity, the conditions under which the activity is performed, and the presence of others during the activity. The chapter outlines how ActiveQoL will be implemented using wearable sensors and smartphones to automatically estimate activity and situational factors. Additionally, it discusses the use of surveys and data collection methods to obtain meaningful data. ActiveQoL enables the evaluation of urban programs by measuring individual QoL and aggregating the data across the population. The article provides examples of ActiveQoL's application in evaluating exercise and sports activities, as well as programs aimed at elderly individuals. These examples highlight the benefits of incorporating ActiveQoL into urban and policy evaluations, such as quantifying the impact of soft and hard programs and tailoring evaluations to specific target groups. The chapter concludes by emphasizing the importance of having a clear vision for the city and conducting objective evaluations to create a happy and distinctive community.

Keywords QoL-based assessment · ActiveQoL · Quality of Life (QoL) · Measuring QoL · Satisfaction level

K. Suzuki
Environment & Energy Innovation Center, Research & Development Group, Hitachi, Ltd., Tokyo, Japan
e-mail: kei.suzuki.zt@hitachi.com

M. Tai (✉)
Systems Innovation Center, Research & Development Group, Hitachi, Ltd., Tokyo, Japan
e-mail: mitsuharu.tai.wu@hitachi.com

T. Sasao
Faculty of Engineering, Reitaku University, Chiba, Japan
e-mail: tsasao@reitaku-u.ac.jp

© The Author(s) 2025
Hitachi-UTokyo Laboratory (H-UTokyo Lab.), *The Architecture of "Society 5.0"*,
https://doi.org/10.1007/978-981-96-2929-9_8

8.1 Background

8.1.1 Assessing Smart Cities from a Citizen's Perspective

For national and local governments and the other organizations that run cities (the "implementer"), the ultimate goal is to create a society in which residents are happy. The goal aligns with the Society 5.0 vision of a human-centric society and the vision for a life of happiness and well-being set out in the "digital garden city nation" strategy promoted by Digital Agency (2024), the government's nucleus of Japan's smart cities. But how close are our communities to realizing these visions? Cities might deliver innovative projects for residents under the name "super city" or "smart city," but such projects can only truly be considered successes if they deliver well-being to residents. When evaluating a smart city, we need to measure how far the project has attained the goal of creating a society in which residents are happy and what means and measures it has taken in running the city.

In reality, though, smart-city projects have usually been evaluated from other angles; evaluations have focused on the extent of the action taken by the party running the city, or have used quantitative metrics to measure the program outcomes. For example, MLIT uses the following physical measures: number of environmental measures, amount spent subsidizing exercise and recreation, number of hospitals, amenity capacity, number of public parks, percentage of green space (by surface area), and percentage of visible greenery (Ministry of Land, Infrastructure, Transport and Tourism 2024). Such measures represent the perspective of the implementer.

However, the extent to which a given program achieves its goal depends on the values and living environment of residents at the time. Hence, it is essential to include metrics that represent the perspective of citizens—metrics that measure how residents perceive the measure, how they perceive the outcomes of the measure, and what outcomes have been welcomed by residents. If we could measure residents' quality of life directly, metrics would be available for measuring the substantial effect of a project—how satisfied people are with everyday life and whether the project has genuinely improved people's lives. However, is measuring residents' happiness directly really possible?

To measure whether residents feel they are satisfied with their lives and whether they feel they are living an independent lifestyle, we would be measuring their quality of life (QoL). WHO defines QoL as "an individual's perception of their condition of life in the context of the culture and value systems in which they live and in relation to their goals, expectations, standards and concerns (WHO 2012)." This definition suggests that QoL can be a measure of the extent to which an individual feels they are affording an independent lifestyle and whether they are finding happiness in their life. On the premise that insights can be gained into the person's perception of their life from their physical health, mental health, level of independence, social relationships, living environment, and from their spiritual or religious background and beliefs; a QoL scale is then defined and evaluated from such background information. QoL was originally a health concept. It was introduced to address a shortcoming of medical scales that only measure the severity of a disease or physical condition; such scales do not fully take into account how a disease or disability can compositely interact with other factors to affect the person's life. The background factors considered in QoL are closely related to the society or urban environment in which the person lives. That is, QoL is an approach to understanding whether residents are happy with the community they live in, making it an important metric for considering happiness when directly measuring residents' satisfaction with a smart city.

8.1.2 Recent Developments, Methods for Measuring QoL

A number of attempts have already been made to directly measure residents' happiness and incorporate the results into policy.

Since 2019, the government has conducted a survey on satisfaction and QoL every other year. The survey is sent online to Japanese citizens selected for sampling (Cabinet Office 2023). The purpose is to understand the multiple facets of Japan's socioeconomic structure from the perspective of people's satisfaction and to apply the findings in government strategy. The results are publicized in a dashboard format, enabling people to analyze from multiple angles the relationship between Japanese citizens' lives and their satisfaction with them.

Another example is in Arakawa City in Tokyo. In 2006, Arakawa started exploring a possible metric for measuring residents' happiness and developed an index named Gross Arakawa Happiness (GAH) (RILAC 2015). Based on this index, the municipality uses a questionnaire for residents. The questionnaire consists of 39 factors across 6 subscales related to the municipality's policies. It gives a total score describing the respondent's happiness level. The response data are analyzed to identify issues, evaluate policies, and decide key strategies for the municipality.

In both examples, paper or online surveys are used to collect the data. The problem with surveys is that one must limit the questions (because the more questions there are, the more stress respondents will experience) and you cannot conduct them often.

As emphasized in Society 5.0 and smart cities, values are becoming more diverse, and society and the environment are evolving rapidly, leading to fluctuations in everyday life satisfaction levels. Therefore, shortening the cycle of policy delivery, evaluation, and feedback is necessary to enable more granular policies that will maximize the level of satisfaction of the times. To that end, policymakers advocate the necessity of a data-acquiring technology and an urban operating system for digital platforms, which will comprehensively manage and digitize urban data. In view of this, general improvement is under way with the government initiative (Cabinet Office 2020). In the case of QoL-based assessment too, defining the index afresh is necessary so that sensing data and other digital data can be used for continuous and real-time tracking of QoL levels. This will ensure that QoL-based assessment can be implemented to be utilized for policies Society 5.0 and smart cities should propose.

In view of these issues, and in view of developments in digital technologies designed to address these issues, we have proposed ActiveQoL as a new instrument for QoL-based assessment (H-UTokyo Lab 2021). ActiveQoL is described below.

8.2   Instrument We Propose: ActiveQoL

8.2.1   An Evaluation Approach Focused on Quality of Activities

Gaining greater satisfaction in the activities we undertake in daily life is a particularly important part of improving QoL and boosting happiness. Therefore, we have proposed ActiveQoL as a QoL-based assessment. ActiveQoL measures an individual's subjective satisfaction in their everyday activities, which include sleep, work, mealtimes, housework, parenting, and leisure. The idea is that when smart-city policies address problems that occur in the abovementioned everyday activities or when it enables people to engage in the activities more comfortably and independently, the level of their satisfaction with such activities will increase, suggesting that one can judge that the effect of the policies has reached residents.

ActiveQoL is designed to allow estimating satisfaction with everyday activities based on four factors, which are believed to significantly shape an individual's subjective satisfaction.

The first factor is the person's fondness for the activity in question. An individual's disposition toward an action will significantly shape their level of satisfaction with it. If the person likes or dislikes the activity unconditionally, then their satisfaction with the activity will be scarcely affected by the activity-specific conditions mentioned below. The second factor is the time the person spends engaging in the activity and the extent to which this time accords with what the person considers desirable. For some activities, a shorter duration will be associated with a higher level of satisfaction, whereas for other activities, a longer duration will be associated with it. An optimum duration exists for some cases, and the person's satisfaction falls when there is a sizable discrepancy between the optimum duration and the actual length of time. In terms of policymaking implications, this factor can vary depending on travel and assistance policies, among other things. The third factor is the conditions under which the activity is undertaken and the extent to which these conditions align with what the person considers ideal. These conditions include the place where the activity is conducted and the means for conducting the activity. Satisfaction in a given activity can vary widely depending on these conditions. Take, for example, the activity of traveling. The person might prefer to use a car so they can get to the location quickly and comfortably. Alternatively, they might prefer to go by bicycle or by foot so they can take in the scenery. Other conditions in this case would include whether the person is traveling in the neighborhood or near their workplace. As these examples illustrate, the person's values and tastes have a significant bearing on their satisfaction with the activity. As these conditions depend on the situational context of the activity, the third factor is affected significantly by policy. The fourth factor is whom the person conducts the activity with and the extent to which the activities coincide with what the person considers desirable. Suppose that the activity is having a meal. Suppose the person engages in other activities under the same condition. We should know from personal experience that our satisfaction with the eating activity would differ markedly depending on whether we are eating alone, with our family, or with a friend. On the premise that these four factors predict satisfaction with everyday activities, we postulated ActiveQoL as the sum of these factors.

When defined this way, ActiveQoL increases when the situational factors of an activity align with the individual's preferences. In this context, favorable and appropriate situational factors mean the person's individual preference including time, place, and person or people to conduct the activity with, and this varies from person to person. ActiveQoL takes into account diversity in values; it takes into account how personal tastes can cause satisfaction to vary for a given activity conducted in a given location. Never before has there been an urban evaluation instrument that is so sensitive to individual attributes. ActiveQoL has potential to be adopted as a foothold for progress toward a city accommodating diverse values and lifestyles.

8.3 How ActiveQoL Works

8.3.1 ActiveQoL's System for Automatic Estimation

ActiveQoL is determined based on the extent to which the person likes the everyday activity in question, how long they have engaged in it, the context in which they engaged in it, who they engaged in it with, and the extent to which these conditions aligned with the person's preferences. Thus, to determine QoL, we need to know three things: (1) what the person is doing (knowing the everyday activity), (2) the situational factors (knowing the duration of the activity, the setting in which the activity takes place, and whom the person is doing the activity with), and (3) whether the person likes the activity and what situational factors align with the person's preferences.

We are currently working on a technology that automatically estimates the first two of these things (the activity and its situational factors) using wearable sensors or sensors embedded in smartphones. The third item concerns personal preferences and so, generally speaking, we would need to ask the person directly using a questionnaire or other method. We are working on two simple methods for knowing this third item. The method involves using a brief questionnaire, searching for patterns

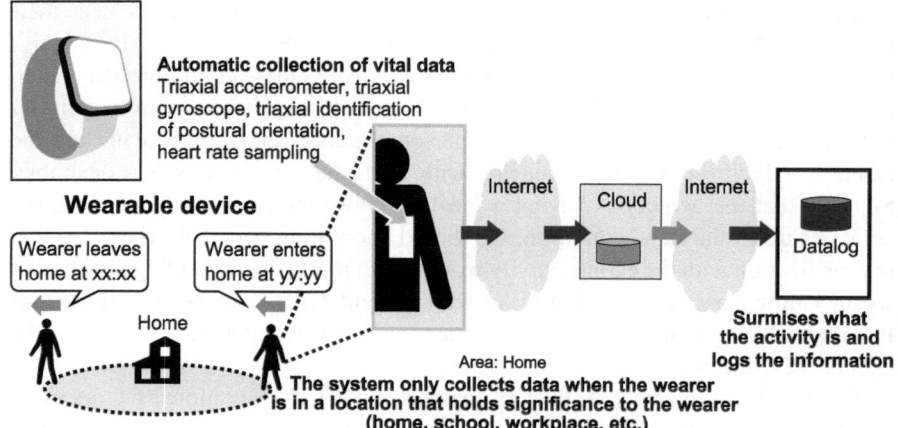

Fig. 8.1 ActiveQoL's system for automatic estimation

in the responses, and then delineating categories. The second method involves repeatedly asking people about their everyday activities, the situational conditions of these activities, and their satisfaction with the activities, to gradually build a body of knowledge about the preferred settings and situational factors for each activity.

The wearable sensors we have in mind are sensors on smart watches, wristbands, and other devices that can be worn similar to everyday clothing and something that has advanced from activity trackers that log the wearer's steps, heart rate, and other activity-related items. Such items are on the market and have started proliferating into general use. They can use GPS data to estimate the person's current location or the person's means of travel, meaning that they can automatically identify when the person is engaging in exercise such as walking, running, or sleeping. Figure 8.1 shows how the ActiveQoL features now under development will make use of wearable sensors and smartphones.

8.3.2 Obtaining Data on Satisfaction with a View to Making ActiveQoL Capable of Automatic Estimation

We eventually want to create a system that can use wearable sensors and smartphones to automatically identify the activity in which the person is engaging. Line is a popular messaging app in Japan. For the first step in this project, we are using a survey on Line, which can send messages at any specified time of the day and input the responses to the messages, allowing us to collect training data along with subjective satisfaction data. Outlined below is our system for conducting a survey via Line and collecting real-time data on the activities users are engaging in, the situational factors related to these activities, and their satisfaction with the activities.

We refer to this system as the ActiveQoL Experience-Sampling System (ActiveQoL-ESM). It semi-automatically samples aspects of participants' everyday lives. At regular intervals, it sends a short set of questions to the smartphones of registered participants. The questions concern the activity the participant is currently engaged in. When a participant notices the message, they send back their responses detailing their current activity. On each occasion, the question set consists of five question items: the type of activity, its time/duration, other people engaging in the activity, the location, and the person's satisfaction with the activity. Because some participants may miss the message or may be too busy to respond, the system is designed so that they only need respond when they notice the message and never feel pressured to respond. The timing of messages, wording of questions, and answer options can be configured differently for each participant, enabling the sender to flexibly customize the sampling content (see Fig. 8.2).

The ActiveQoL-ESM is still in the testing stage, and a number of problems still need to be addressed. For example, the process of responding to questions is still very stressful, and its function, unable to send questions at the point where the activity discontinues, times out. However, QoL-ESM is effective for reducing the investigator's workload; unlike with paper surveys, QoL-ESM requires no extra process to digitize the response data, and the question and answer options format can be changed on the fly. It can also reduce the workload for respondents; if respondents

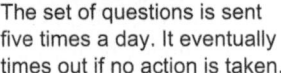
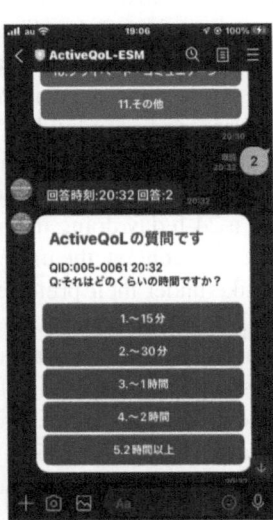
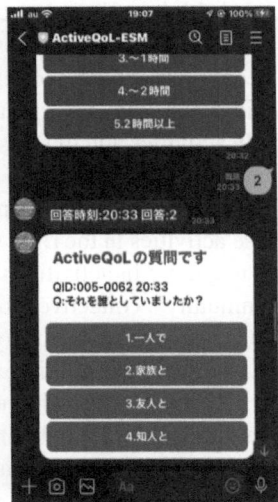

The set of questions is sent five times a day. It eventually times out if no action is taken.

When the person answers one question, the next ones are sent sequentially.

Each sequence has four to five questions.

Fig. 8.2 Collecting data about activities through line

have a smartphone, they can simply tap the relevant option. We will further alleviate the burden on respondents by reducing the number of occasions they have to respond and using smartphone sensor data.

8.4   Using ActiveQoL to Evaluate an Urban Program

The idea with ActiveQoL is that it is used to measure an individual's quality of life in relation to a particular activity of such individual and then the aggregated data is used to evaluate the effectiveness of a policy. The sample of individuals can be set as desired according to the sample contents to be evaluated. If a policy is to be evaluated and delivered to all residents as a whole, there would be no need to restrict the sample, but supposing the policy of interest is a policy for elderly people and its effectiveness is to be evaluated among this demographic, then the sample would be restricted to elderly residents. By narrowing down the sample contents of the policy in question and the attributes of the intended audience of the policy, the results will indicate more directly how effective the policy is relative to its intended audience and goal, giving better insights into how the policy could be tweaked and improved. This section discusses how sampling data collected by ActiveQoL can be used to evaluate smart cities.

ActiveQoL gives an activity satisfaction score on a 5-point Likert scale (with 1 indicating unsatisfied and 5 indicating satisfied). On the premise that an individual's QoL is higher when they engage in activities they like and do so in the right situational context, and supposing that an individual engages in five activities a day (sleep, work, mealtime, exercise, and housework), then ActiveQoL would be as valid as the daily total or daily average of an individual's satisfaction with those five daily activities. On this basis, one focuses on the situational conditions that apply in those activities the person says they like. As an individual's QoL will be higher if the person spends more time in a 24-h day doing the activities they like and doing these activities in the right situational context, the more people in a community who are engaging in activities they like under their preferred conditions, the higher the community's collective QoL will be.

Consider this example of using ActiveQoL to evaluate a smart city. In this example, we are interested in exercise and sport activities that residents engage in. We start by finding out whether the individuals like exercise and sport. We then use wearable sensors to obtain data on their satisfaction level upon engaging in the activity. This response data can be inputted into the categories in the table shown in Fig. 8.3. Into (1), we input whether the person said they feel satisfied with the activity regardless of their disposition toward the activity itself. Into (2), we input whether they said that they do like exercise and sport but are nonetheless not engaging in this activity or are engaging but are dissatisfied with it. Into (3), we input whether they said that they dislike exercise and sport and are either not engaging in such or feel unsatisfied with it. In this way, we assign each respondent to one of the three categories and then count each category's tally and percentage of the total.

Disposition toward and satisfaction with exercise and sport		Satisfaction level			
		Does not engage in	Unsatisfied, rather unsatisfied	Neither satisfied nor unsatisfied	Satisfied, rather satisfied
Preference	Likes	(2) Likes but unsatisfied = QoL low			(1) Favorable situation = We would expect QoL to be high
	Neither likes nor dislikes				
	Dislikes	(3) Person will not be forced to do something they dislike = In most cases, the QoL will fail to increase Requires different approach from that used in cases where the person is not doing something they should be doing			

Fig. 8.3 How to use ActiveQoL

A smart city is judged by the percentage of the sample in (2). This category represents the people who are unsatisfied or unengaged despite liking the activity in question, and the presence of such people poses a problem. If people feel dissatisfied with an activity they would normally enjoy, it suggests that improvements are needed. Thus, to get an idea of the status quo, we aggregate respondents' data and sort respondents into categories to determine whether there are lots of people, or only a few people, for whom the activity needs to be improved. We can then consider an action to take for each category. In this way, the present whereabouts of the smart city can be understood relative to a given activity and an appropriate action to take for people who need the activity to be improved can be considered.

Once the action has been taken, it is possible to see whether people have moved from (2) to (1). If the population of (1) has increased, it implies that the action was successful.

As equipping every single resident with wearable sensors is probably unfeasible, satisfaction would be measured in just a sample of residents. In this case, the sample's average QoL is obtained and the distribution analyzed or the average between different geographic zones or between age groups is compared. QoL trends might also be tracked longitudinally (by seeing how QoL changes over time).

The sample subjects would have to be narrowed down, if a local government wants to use ActiveQoL score to guide its policymaking or evaluate its policies. The local government might be interested in a particular infrastructural asset such as a public park (including the park's amenities and roads) or in evaluating whether a service it delivers to elderly people is having a positive effect in elderly people's daily lives. In this case, the sample would have to be narrowed down to the daily-life activities linked to the target location or the subjects would have to be narrowed down to users of the service.

ActiveQoL, despite being a service for data provisions in view of service evaluation, could potentially be applied to digital services, if the service in question is designed to change people's behavior. If ActiveQoL can evaluate commercial services and urban policies, then it can potentially have a wide range of usage.

If ActiveQoL is to be used by local governments, then it could serve as a KPI for evidence-based policymaking (Cabinet Office 2024), which the Cabinet Office is encouraging the government's departments, agencies, and local governments to adopt. Evidence-based policymaking is defined as an initiative that clarifies the basic framework for the policy by clarifying the objectives of a policy, the logical connection between the policy's means and ends, and finding as much evidence as possible data or other forms of supporting evidence for this connection. When policies are developed and implemented in this way, they will make possible more minute, swift, and accurate policy formulation and deliveries, and visualize the effect of the policy to enable any stakeholders to appreciate it. The Cabinet Office recommends that when adopting evidence-based policymaking, policymakers should use a logic model (Newcomer et al. 2015) to set KPIs and run a plan-do-check-act cycle. In the basic logic model, the first component is resources. In the resources field, the human, infrastructural, and financial resources, and other inputs are listed. These are the resources and stakeholders required to implement the program. Having clarified this information, we can start on the other components of the logic model: we envisage the long-term outcomes, intermediate outputs, and short-term outcomes, and then decide on the activities required to produce the short-term outcomes and what the immediate outputs of these activities will be. ActiveQoL can help in this process by measuring the immediate outcomes. Consider the example in Fig. 8.4. In this example, a local government is using a logic model to plan a program for encouraging elderly people to engage in the community. To measure the expected effect of this policy in changing behavior, the local government has set a number of KPIs, one of which measures QoL stemming from behaviors related to communication and travel (walking or having diverse travel destinations).

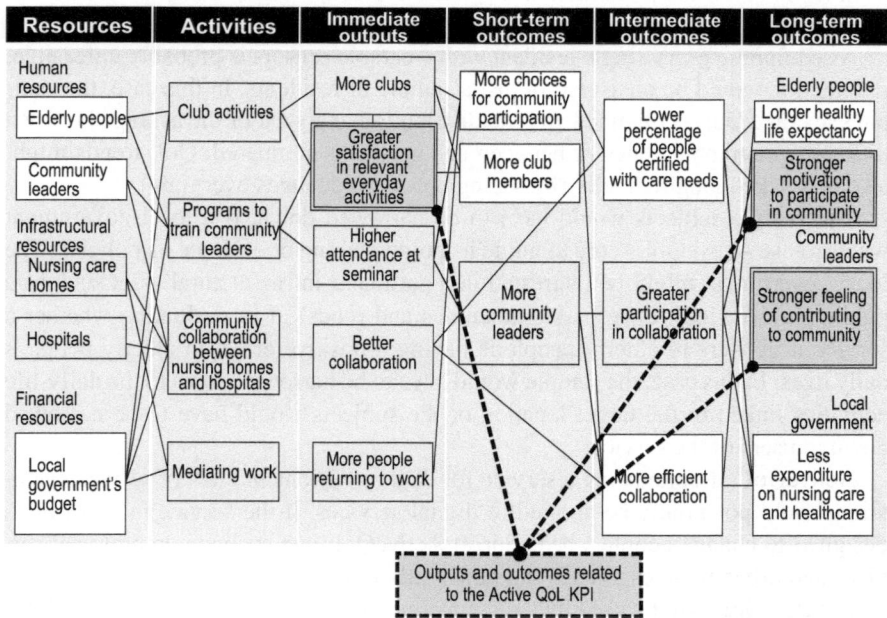

Fig. 8.4 Example of a logic model for a program designed to encourage elderly people to engage more in the community

8.5 Demonstrating the ActiveQoL Concept

To test the ActiveQoL concept, we applied ActiveQoL to a program run by Kakogawa City, a municipality in Hyogo Prefecture. The program involved running a seminar for elderly people who were novices with smartphones. The seminar was designed to help elderly participants master the use of their smartphones, but the longer-term goal was to give the participants opportunities for communication and raise their motivation to go out and try new things. The program's success was judged by whether changes occurred between before and after the seminar in the participants' everyday activities and satisfaction with such. These changes could be visually represented with a radar chart indicating satisfaction with each everyday activity (Fig. 8.5). Prioritizing everyday activities with a relatively strong impact on QoL, the program team uses a plan-do-check-act cycle to evaluate improvements in the program, evaluate the program's connections with other programs, devise how to make the program more effective, and see how it might integrate with other programs.

After the seminar, the program team used our ActiveQoL-ESM tool to ask participants about their everyday activities and their satisfaction with these. However, because the participants had no experience with smartphones prior to the seminar, the initial data was collected by means of a paper survey. As both the paper survey and questions on the ActiveQoL-ESM encumber the participants, the automatic

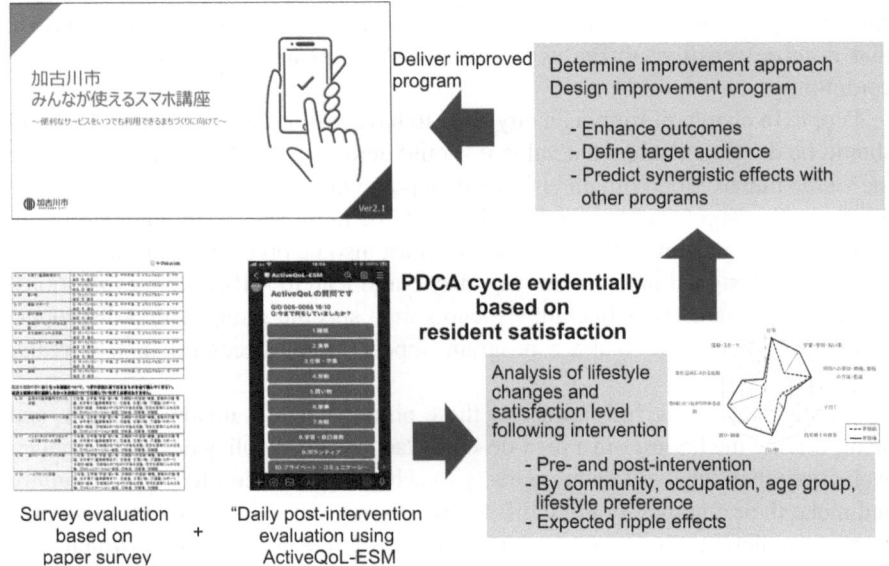

Fig. 8.5 Demonstration of ActiveQoL in evaluating a program for elderly people (run with Kakogawa City, Hyogo Prefecture)

data-collection system (as shown in Fig. 8.1) must be urgently developed. The plan now is to analyze the collected data to identify how the seminar can be improved and identify other insights for developing effective policies for elderly people.

8.6 Toward Urban QoL-Based Assessment

A key benefit of incorporating ActiveQoL into evaluations of urban programs is that it can offer evidence to demonstrate whether an expenditure on an activity is cost effective, particularly because it allows the outcomes of soft programs (programs involving intangible services such as fitness and wellness programs) to be quantified; until now these have been difficult to evaluate. ActiveQoL is also useful for hard programs (programs involving tangible infrastructure). Once the infrastructure has been built, that is not the end of the story; what matters is how the infrastructure operates and whether it delivers value to residents. There has been a dearth of techniques for evaluating completed infrastructure projects. ActiveQoL can help address this problem by quantitatively evaluating how well a piece of infrastructure is working.

Another advantage of ActiveQoL is that it allows policymakers to configure a QoL metric by narrowing down only to the target traits. Policymakers could, for example, narrow down the evaluation to mothers raising children, to residents who use wheelchairs, or to other demographics that are often overlooked when policymakers rely on an average score among all residents. Such customization can help yield insights that can contribute toward community development that is more sensitive to the needs and problems that affect a minority of the community.

People in charge of running a city need to have vision for the city (how the city should be developed and what value it should deliver to residents) and to evaluate the vision impartially. With no vision, urban policymakers cannot identify with any precision the target groups whose QoL should be improved, resulting naturally in average, homogeneous cities. Evaluations must also be objective and impartial. If surveys are designed in a certain way, they could lead to self-serving evaluations, in which the evaluator deliberately focuses on a specific group with a disproportionately high QoL to make a program appear more successful than it really has been.

If policymakers are fully aware of these pitfalls when evaluating the QoL and incorporating the findings in a plan-do-check-act cycle for policy design and policy evaluation, then they will be more likely to achieve their vision for the community and make their community more distinctive. We urge those involved in urban programs to remember that their ultimate purpose is to create a society in which residents are happy and to apply ActiveQoL effectively for this purpose.

References

Cabinet Office (2020) *SIP saibā ākitekucha kōchiku oyobi jisshō kenkyū no seika kōhyō* [SIP cyber architecture building and empirical research results public presentation]. https://www8.cao.go.jp/cstp/stmain/20200318siparchitecture.html. Accessed March 30, 2024

Cabinet Office (2023) *Manzokudo seikatsu no shitsu ni kansuru chōsa* [Survey on satisfaction and life quality]. https://www5.cao.go.jp/keizai2/wellbeing/manzoku/index.html. Accessed March 30, 2024

Cabinet Office (2024) *Naikakufu ni okeru EBPM e no torikumi* [Cabinet Office's initiatives for EBPM]. https://www.cao.go.jp/others/kichou/ebpm/ebpm.html. Accessed March 30, 2024

Digital Agency (2024) *Dijitaru denen toshi kokka kōsō* [Digital garden city nation]. https://www.digital.go.jp/policies/digital_garden_city_nation/. Accessed March 30, 2024

H-UTokyo Lab (2021) *Hito chūshin no sumātoshiti no hyōka to QoL* [People-centric smart city evaluation and QoL]. In: Third Habitat Innovation forum, October 2021. http://www.ht-lab.ducr.u-tokyo.ac.jp/wp-content/uploads/2021/10/5b6a7c4315a29b81340890b7402209ff.pdf. Accessed March 30, 2024

Ministry of Land, Infrastructure, Transport and Tourism (2024) *Toshi monitaringu shīto* [Urban monitoring sheet]. https://www.mlit.go.jp/toshi/tosiko/toshi_tosiko_tk_000035.html. Accessed March 30, 2024

Newcomer, K.E., Hatry, H.P. & Wholey, J.S., eds. (2015) Handbook of Practical Program Evaluation, 4th Edition, San Francisco, CA: Jossey-Bass & Pfeiffer Imprints, Wiley.

Research Institute for Local government by Arakawa City (RILAC) (2015) Gross Arakawa Happiness (GAH). https://rilac.or.jp/?page_id=307. Accessed March 30, 2024

World Health Organization (WHO) (2012) WHOQOL: Measuring quality of life. https://www.who.int/toolkits/whoqol. Accessed March 30, 2024

Chapter 9
Human Resource Development for Smart Cities

Shin Osaki and Atsushi Deguchi

Abstract Smart cities require human resources for building and operating a system that covers a variety of specialized fields. These human resources can be divided into three types: architects, coordinators, and collaborative specialists. Architects draw out an overall vision of the smart city, lead the coordination of stakeholders, and oversee the overall project. Coordinators work under architects and ensure that the collaborative specialists work together to materialize the desired mechanisms and projects.

Programs that develop such human resources apply both an analysis approach (to provide broad knowledge on smart cities) and a synthesis approach (to develop the ability to integrate knowledge and create plans).

It is important to advance human resource development as well as provide graduates with appropriate work. From this perspective, human resource development and smart city organizational structure creation are closely intertwined. This requires not only the training and placement of experts but also education for improving citizen literacy and living laboratories as forums to collaborate with citizens.

Keywords Human resource development · Architect · Coordinator · Collaborative specialist · Analysis and synthesis approach

S. Osaki (✉)
Neighverse Inc., Tokyo, Japan

Sustainable Society Design Center, Graduate School of Frontier Sciences, The University of Tokyo, Tokyo, Japan
e-mail: osaki@edu.k.u-tokyo.ac.jp

A. Deguchi
Department of Socio-Cultural Environmental Studies, Graduate School of Frontier Sciences, The University of Tokyo, Tokyo, Japan
e-mail: deguchi@edu.k.u-tokyo.ac.jp

Hitachi-UTokyo Laboratory (H-UTokyo Lab.), *The Architecture of "Society 5.0"*, https://doi.org/10.1007/978-981-96-2929-9_9

9.1 The Human Resources Required for Smart Cities

9.1.1 Cross-Field Jobs for Smart Cities

Smart cities involve programs that are grounded in digital technology and that span multiple fields, including mobility, environment/energy, disaster management/crime prevention, wellness/healthcare, and community (Fig. 9.1). In some cases, digital technology is integrated with a single field, such as combining simulation with transportation or combining AI with crime prevention. In other cases, digital technology is integrated with multiple fields. Mobility as a service (MaaS), for example, involves integrating data between transportation, tourism, and account settlement.

The above table illustrates how smart cities involve the provision of services that integrate multiple fields. It also illustrates how these services rely on a digital foundation or platform consisting of an urban operating system, a system for linking data across multiple fields, and data governance.

Major category	Field	Program theme
Service	Mobility	Transportatoin/mobility, logistics, transport hubs
	Environment/energy	Environment, energy, aquatic resources, waste
	Disaster management/crime prevention	Disaster management, crime prevention
	Infrastructure/amenities	Infrastructure maintenance, urban planning, amenity management, housing, construction, real-estate
	Wellness/healthcare	Wellness, healthcare, nursing care
	Industry/economy	Agriculture/forestry/fishers, tourism, local economic development, industry creation, academic-industrial collaboration, digital currency/payment, workstyles
	Community	Local community building, community self-governance, social activities
	Education/culture	Education, parenting, culture/art
	Government/administration	e-services, digital operation, security
Foundation/platform	Digital platform	Urban operating system, data-linkage platform, digital communication network, open data, 3D urban model, data governance, accessibility
	Organizational framework	Public-private partnership, public involvement, transparent management, monetization
	Human resources	Training and deploying smart-city human resources, programs for addressing digital divide

Fig. 9.1 Smart-city fields. (Counselor to the Cabinet Office Director-General for Economic, Fiscal, and Social Structure, Cabinet Office Secretariat for Science, Technology, and Innovation Policy 2022)

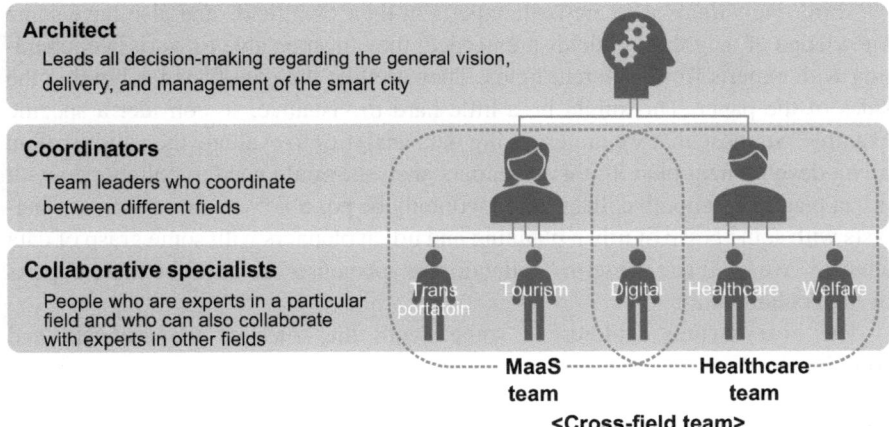

Fig. 9.2 Three types of human resources required for smart cities

9.1.2 Three Types of Human Resource Required for Smart Cities

Given that smart cities span multiple fields and also require digital literacy, vertically structured organizational silos are inadequate for coping/responding. Rather, a cross-field approach and the human resources to coordinate activities are required. Figure 9.2 shows the three types of required human resources.

First is the human resource to provide the overall leadership and direction for a smart city. The human resource here sees the big picture and directs all the decision-making delivering the vision. Their role is to build smart-city architecture with a comprehensive understanding of the areas covering the foundations/platforms (such as an urban operating system) to the services. The Cabinet Office has dubbed this role as that of the "architect.[1]"

Next are the team leaders who coordinate activities between the fields. Inter-field teams are necessary because smart-city services involve a combination of different fields. However, these teams will never operate effectively just by gathering members specializing in particular fields. That is why people are needed who can play a coordinating role, mediating between the different fields—namely, "coordinators."

---

[1] The Cabinet Office has defined the architect's role as that of "masterminding the super city project as a whole, including identifying the local needs/objectives, drawing up a business plan, and deploying technological innovation." This definition was mentioned in the terms for applying for super city status, denoting that having an architect is a prerequisite for designation. The definition does not, however, imply that the architect must always be the same party, given that the Cabinet Office also included the following proviso: "When efforts move from the planning stage to the implementation stage, the architect may need to be changed (Cabinet Office Secretariat for Promotion of Regional Revitalization 2020)."

Third, individuals who are both experts in their own fields and also have some knowledge of neighboring fields are needed; they must be able to work in cooperation with experts from different fields. They are like the glue flaps for binding the sides of the paper. This might be a little hard to visualize, so consider a specific example. Suppose that we are using big data analytics to evaluate the validity of an urban development plan. In this case, data analysts would need to collaborate with urban planners. Smooth collaboration will only be possible when we have data analysts with some grasp of urban planning and urban planners with some grasp of data analysis. We shall use the term "collaborative specialists" to describe these experts in such collaboration.

The next section explores in some depth the roles of the architect and coordinators.

9.2   The Roles of the Architect and Coordinators

9.2.1   The Role of the Architect

The architect is in overall charge of the smart city (or super city). The Cabinet Office has defined the architect's role as that of "masterminding the super city project as a whole, including identifying the local needs/objectives, drawing up a business plan, and deploying technological innovation (Cabinet Office Secretariat for Promotion of Regional Revitalization 2020)." This means that the architect manages all the processes in Fig. 9.3 (which is based on an illustration that appears in SIP's Smart City Reference Architecture White Paper) (Cabinet Office 2023). The architect performs two processes simultaneously: In one direction in the figure, the architect identifies the community's needs and then sets the strategic vision, defining objectives, setting key goal indicators and key performance indicators, and then steering the management of the smart city in a direction aligned with this strategy. In the other direction in the figure, the architect uses data provided by the smart-city's assets to plan, deliver, and operate an urban operating system consisting of the platforms for delivering the services. Thus, the architect has an overall grasp on both the physical and digital tracks and manages the overall delivery of smart-city services by coordinating the activities of the stakeholders.

For this role, the architect must be proficient in a wide range of fields. The challenge lies in the scarcity of individuals possessing such extensive expertise. The architect, therefore, can engage a small team of experts who complement each other's expertise to cover a wide range of specialized fields.

To deliver the vision for a smart city, collaboration and coordination among different stakeholders are necessary. Thus, another key role of the architect is to lead such collaboration. The architect must, on the one hand, persuade legislative and administrative bodies (or the redeveloper in the case that the smart city is being created in an area for redevelopment) of the merits of the smart city and make the

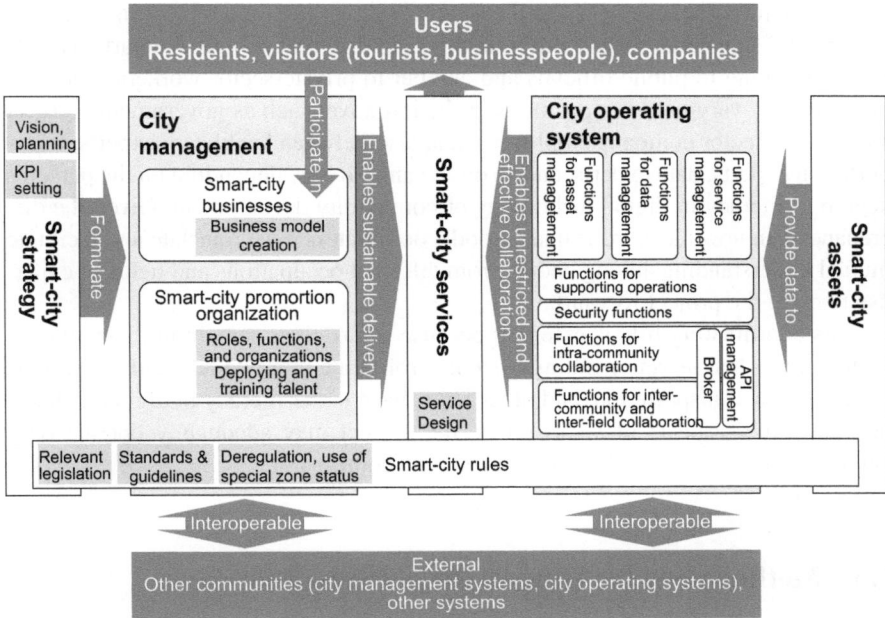

Fig. 9.3 General picture of the smart city reference architecture. (Figure adapted from (Cabinet Office 2023))

arrangements for these bodies to collaborate in the project. On the other hand, the architect must also bring on board the private-sector stakeholders who have the necessary technology and resources to deliver the vision for the community. To meet both of these requirements, the architect must arrange a system that will make the smart city a viable business proposition and something that serves the public interest. In some cases, the architect might need to consider creating a collaborative framework whereby local universities and research institutions are granted access to research fields, contingent upon their reciprocal contribution of specialized knowledge and research outcomes to the community.

9.2.2 The Role of Coordinators

The role of the coordinators, as set out below, is our own conception of what the role should be, as the government has not defined this role. Coordinators work with experts in multiple fields to enact the architect's vision for the smart city. As such, coordinators must fully understand the architect's complete vision and its essence and then effectively coordinate with experts in various fields to shape the system and its services.

The coordinators can be likened to translators. At the time of writing, the term "smart city" can mean several different things to different people. A smart city may mean one thing to public officials and another to private-sector workers. The connotation may vary between positions in the initiative, such as government workers and private-sector company employees, and also between healthcare experts, transportation experts, data experts, and so on. It can also vary depending on the person's degree of literacy in digital technology or community development. Given the discrepancies in how the word is understood, coordinators must translate and facilitate mutual understanding among people from different occupations and fields of expertise to drive the project forward.

This coordinating role is essential because smart cities remain in their infancy. Eventually, when urban operating systems proliferate more widely and the experts in healthcare, transportation, and other fields have better literacy in digital technology and other aspects of smart cities, the project may adequately operate only among a team of specialists without needing a coordinator.

9.3   Methodology for Human Resource Development

9.3.1   A Three-Tier Pyramid

How, then, do you develop the human resources for the smart city, particularly the people who will take on the architect and coordinator roles? Granted, this question may have no single correct answer. However, we have set out the following approach for gaining insights into a possible methodology for human resource development.

I just mentioned that collaboration may eventually be possible without coordinators once the experts have acquired knowledge and proficiency in digital technology and related fields. Once we have a solid team of specialists with their glue-flap function to collaborate with, they will be able to spontaneously play a coordinating role. If so, then it would imply that specialists working in collaboration are also budding coordinators. By the same token, once coordinators start to master the ability to set and execute an overall vision for a smart city, they can start discharging the role of the architect; a coordinator is therefore a budding architect.

We can therefore see the three roles—collaborative specialists, coordinators, and architects—as a three-tier "evolutionary" pyramid through which experts in a particular field ascend. They ascend by collaborating with experts in other fields and gaining, from such collaboration, the strategic skills (strategic planning, big-picture thinking, and execution) to implement the smart city as a whole. In this model of human resource development, we would devise programs designed to help people reach the next level in the pyramid.

9.3.2   Analysis and Synthesis

No one has yet established an actual program for training the coordinators and architects. Neither do we have a clear idea of how these two programs would differ. However, it would probably be effective in both programs to combine analytical and synthetic approaches.

An analytical approach dissects the subject matter in question, breaking it down into constituent components to enable an easier understanding. A synthetic approach explores and combines (synthesizes) different ideas and knowledge about the subject matter to generate a logical solution. If a program takes an analytical approach, it might involve a lecture-style program in which attendees study the specialist theories, ideas, and knowledge related to smart cities. If a program takes a synthetic approach, it might involve applying these specialist theories, ideas, and knowledge to create a hypothetical scenario for building a smart city.

With an analytical approach, the important thing is to master contemporary problems and initiatives across a wide variety of fields related (or potentially related) to smart cities. The key point here is that we are training generalists; rather than having attendees delve into a particular field, we would have them gain a broad understanding of the key technologies and themes spanning multiple fields.

With a synthetic approach, the important thing is to master the ability to formulate a theoretically coherent plan. Specifically, we would focus on three target competencies specific to smart cities. These three competencies are described below.

The first is the ability to engage in demand-side thinking. Whereas supply-side thinking focuses on what you could achieve with the technology, demand-side thinking focuses on what the community needs. The purpose of creating a smart city is to achieve the vision for the community and address the community's needs, not to build a smart city for its own sake.

The second is the ability to formulate an audacious breakthrough strategy. Instead of an elegant all-around strategy that tries to leave nothing out and offend no one, it is much better to focus on programs that businesses can get on board with and those that will reach the heart of what residents want. In other words, you need ideas for getting things moving and ideas that will run by themselves.

Third is the documentary skill—the ability to document information in a format that is clear and accessible to stakeholders of different positions and interests. I previously mentioned that coordinators are translators. Documentary skills, then, are translation skills. Specifically, these are the skills needed to put together a coherent action plan for the smart city. We need people who understand the community's strengths and needs and can articulate in plain language how the adoption of smart-city technologies could create new value for the community.

9.4   Training for Both Experts and Laypeople

9.4.1   Applying a Human Resource Development Program and Building the Organizational Infrastructure to Deploy Human Resources

So far in this chapter, I have presented three types of human resources needed for smart cities—architect, coordinators, and collaborative specialists—and presented a model showing these roles as tiers on a pyramid. I then discussed two approaches to human resource development: an analytical approach, which would involve a lecture-style program, and a synthetic approach, which would use exercise-based learning.

However, smart-city developers and local governments are yet to create such a human resource development program. We therefore considered the possibility of utilizing learning opportunities provided by universities and other external parties, including seminars and other programs for recurrent education and reskilling.[2]

The value of such external programs lies not just in the educational content itself but in the networking opportunities they offer attendees. Attendees take a step outside their usual business milieu and work and think alongside fellow attendees, and this experience encourages the formation of networks in which everyone is on the same level. By gaining connections in other fields, other companies, and other cities, people will share information with each other and continually learn from one another. This approach will strengthen each person's abilities, and the synchronistic drive to keep learning from one another will be crucial in creating a high-quality workforce for the neighborhood of a smart city as a whole.

In conjunction with efforts to train expert human resources, the local government should capitalize on the outcomes of human resource development by establishing a robust organizational framework, including creating new positions within government administration that span different fields and establishing a division in charge of smart-city advancement that involves overlapping fields. It would be a waste to train up expert human resources but have no positions to deploy the human resources in.

---

[2] One example of a human resource development program relevant to this chapter is the Smart City School of the University of Tokyo's Graduate School of Frontier Sciences. This school uses both analytical and synthetic approaches, with a curriculum involving lectures, round-table discussions, site visits/experiences, and task-based learning. The school was launched in April 2022. It has a spring course and an autumn course. The school head is Atsushi Deguchi, from the University of Tokyo. https://smartcity-school.k.u-tokyo.ac.jp. In another example, the Smart City Institute has started looking into a possible program for training architects, and the progress of its planning is accessible for free on its website, which also includes many freely accessible video resources. https://www.sci-japan.or.jp

9.4.2 *Raising Citizens' Literacy and Training Leaders*

In a time of active flows of personal data on social media, it is also important to improve public literacy in smart cities—to encourage a correct understanding of digital technology and data among the public at large. It is especially important to do so among the target community of a smart-city project in terms of the construction and development of the smart city. Data plays a central role in efforts to run and improve a smart city, and when residents know how to use data properly, the smart city can be run in a more effective and inclusive way.

One way of improving the public's data literacy is to enhance citizen education programs. Local educational organizations and community centers can provide opportunities for the public to learn the basics of interpreting data, learn data analytics techniques, and learn how to read graphs and charts, with detailed case studies illustrating how data applications serve in everyday life and contribute to a better city. You might, for example, have data indicating a rise in reported crimes, but the response would depend on how the data are interpreted. One interpretation could be that crime is on the rise amid a breakdown in law and order; another is that actual crime levels are the same as before, but it is just that with better investigation methods, there are higher arrest rates (more cases are being counted as crimes).

Another effective strategy is to disseminate digital assistance in a location accessible to residents, as in the example of Kashiwa-no-ha Smart City's "digital concierge." By being able to make inquiries about how to use a smart phone or app, residents will be able to gain better data literacy and engage more with community services and apps.

It is also essential to improve public awareness about personal data protection and security. The public needs to know how personal data are collected and used and what the mandatory safeguards are. Workshops and seminars can raise public awareness about these matters by informing attendees about the risks of sharing data and how to address them.

Another strategy is to use data applications to provide real-time updates about the community; this helps in building a smart (digitized or digitally literate) community. The data can be graphically represented in a dashboard format to provide citizens with an intuitive understanding. It is also a good idea to use a dedicated app that informs users about how the data can be graphically represented, interpreted, and applied; this will help empower citizens to be more conscious in their use of data.

Finally, I want to emphasize the importance of citizen participation. When citizens actively participate in urban planning and policymaking, they will share more feedback data digitally, enabling better decision-making. To encourage such participation, government bodies should improve the way they communicate updates on their urban development and community-building efforts. With better communication from the government and better digital literacy among residents, it will be possible to make use of tools for interactive citizen participation.

In summary, critical to the success of a smart city is the task of enhancing citizens' digital literacy. To create a foundation for building a better tomorrow, smart cities must encourage citizens to use data effectively through a combination of approaches, which include education, public awareness raising, visual communication, and campaigns to encourage citizen participation. Citizens' data literacy is a crucial requisite for developing a sustainable smart city, and it also influences citizens' social acceptance to the smart city, making it a critical determinant of success or failure. The point that needs to be emphasized here is that, in addition to developing expert human resources, the local government must also make steady efforts to improve residents' data literacy.

9.5 Organizational Framework

9.5.1 Public-Private/Civic-Academic Consortia

When considering human resource development, one must also consider the organizational framework for it. The framework should involve multi-stakeholder collaboration, with stakeholders serving their respective roles. Rather than just having the private sector involved, the involvement of local government in launching the programs and procedurally managing the smart-city project will be needed, as will the involvement of local universities that can engage long term and contribute their specialized knowledge.

The organizational framework should specify, in a way that reflects the attributes of the target community, the roles of the public-field stakeholder (local government), private-field and civic stakeholders (companies, business operators, residents), and academic stakeholders (universities, research institutes). To enable collaboration between the public-field, private-field, and academic field stakeholders, an organization in charge of managing the project is needed, while the key to bringing local residents on board is a living lab. Figure 9.4 illustrates H-UTokyo Lab's model for an organizational framework for smart cities, which is based on the organizational framework for Kashiwa-no-ha Smart City.

9.5.2 Project Management Organization

At the center of the figure is the project management organization. This organization leads the consortium and manages its projects. A smart city requires coordination across different sectors (data, healthcare, energy, urban planning, and so on), and the project management organization is responsible for such coordination. Therefore, an existing project management organization that manages multiple different project themes as a community-based organization, such as UDCK in Kashiwa-no-ha Smart City or town management organization, is suited for this task.

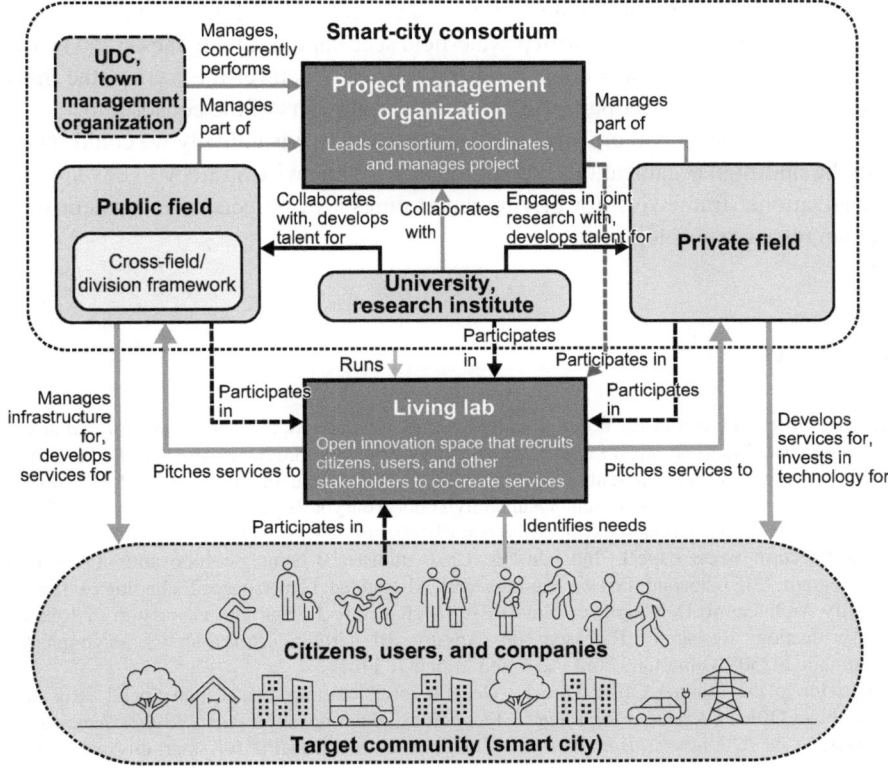

Fig. 9.4 Model of organizational framework for smart cities

If there is no such organization available, then it is worth considering launching one. Additionally, internal project management may be required within the government. In such a case, it is a good idea to recruit external experts to help in breaking down vertical silos cross-sectionally, creating a system for obtaining expert advice, and creating a cross-sectional division.

9.5.3 Living Lab

The living lab in the figure serves as a space for open innovation (see Chap. 7). In this space, citizens can participate in the creation of services, ensuring that services are not delivered unilaterally by a private-field consortium. Remember, the purpose is to avoid a situation where a consortium delivers services to the community from the top to down in favor of a situation in which needs are identified from the bottom up and citizens participate in the creation of the services to address these needs. The living lab should therefore recruit citizens and users, along with other stakeholders, so that it can function effectively as a space for co-creation and open innovation.

It bears repeating that a living lab, as a space for creating new services, is supposed to match citizen needs with private-field seed capital. Living labs could be run by the public, while others could be privately run. Whatever the format, the main issue is to ensure close collaboration with other members of the consortium.

Regarding the overall organizational framework, as it is unlikely we could establish the optimum organizational setup in one go, a stepwise strategy to develop the organizational framework for driving smart cities will be needed in tandem with human resource development.

References

Cabinet Office Secretariat for Promotion of Regional Revitalization (2020) *Sūpāshitigata kokka senryaku tokubetsuku no shitei ni kansuru kōbo yōryō* [Requirements for applying for supercity strategic special zone status]. December 25, 2020, Rev. Edn. February 19, 2021. P.7. https://www.chisou.go.jp/tiiki/kokusentoc/supercity/koubo/youryou01.pdf . Accessed April 15, 2024.

Cabinet Office (2023) *Sumāto Shiti: Rifarensu ākitekucha howaito pēpā* [Smart city: Reference architecture white paper], 2nd edn. In: Cross-ministerial Strategic Innovation Promotion Program (SIP) Second Phase, Big-data and AI-enabled Cyberspace Technologies /Smart City Architecture Development /Smart City Architecture Design and Promotion of Related Verification Research (Released on August 10, 2023). https://www8.cao.go.jp/cstp/stmain/20230810smartcity.html. Accessed March 7, 2024.

Counselor to the Cabinet Office Director-General for Economic, Fiscal, and Social Structure, Cabinet Office Secretariat for Science, Technology, and Innovation Policy (2022) *Sumātoshiti shisaku no KPI settei shishin ni tsuite* [Guidance on setting KPIs for smart-city programs]. (Released on April 2022). https://www8.cao.go.jp/cstp/society5_0/smartcity/01_sc_sihyou.pdf. Accessed April 15, 2024.

Chapter 10
Data Ecosystem

Soichi Furuya and Atsushi Deguchi

Abstract This chapter explores the concept of a data ecosystem, wherein data from various domains, such as community-building, transportation, and energy, is utilized to enhance smart cities. By sharing and interlinking data across multiple sectors, new businesses can emerge, establishing data as a pivotal asset for sustainable urban development. The chapter explores the infrastructure needed to support this data ecosystem, including networks, data storage platforms, and marketplaces. Through real-world examples and case studies, readers will gain insights into how data can be leveraged to improve services, reduce costs, and generate new value in smart cities.

Section 10.1 explores the utilization of data as a new community resource in smart cities. In Sects. 10.2 and 10.3, the concepts of a data ecosystem and its data infrastructure are explained. Section 10.4 outlines various measures and functions designed to promote data use. Section 10.5 discusses the challenges and impediments identified through a study conducted in Kashiwa-no-ha Smart City. Section 10.6 examines how data-driven decision-making processes can enhance urban management and services. Finally, Section 10.7 outlines the future prospects and potential challenges of implementing data ecosystems within smart cities.

Keywords Data infrastructure · Urban operating system · Data marketplace · Online survey · AI camera

S. Furuya (✉)
AI Transformation Strategy, Digital Engineering Business Unit, Hitachi, Ltd., Tokyo, Japan
e-mail: soichi.furuya.xz@hitachi.com

A. Deguchi
Department of Socio-Cultural Environmental Studies, Graduate School of Frontier Sciences, The University of Tokyo, Tokyo, Japan
e-mail: deguchi@edu.k.u-tokyo.ac.jp

Hitachi-UTokyo Laboratory (H-UTokyo Lab.), *The Architecture of "Society 5.0"*, https://doi.org/10.1007/978-981-96-2929-9_10

10.1   Data as a New Community Resource

Society 5.0 is a data-driven society. In several social activities, data will be the basis for decision-making and the operation of infrastructure. Likewise, smart cities generate and use data in many different fields (including community-building, transportation, and energy) with the consent or permission of the parties who have an interest in the data to improve satisfaction, add value, deliver slick services, or cut waste.

When data obtained in the delivery of a service is used on an ongoing basis, it can help generate new businesses in the community, making it a valuable resource for the sustainable operation of a smart city. Crucial to the sustainability of a smart city, then, is making use of data that spans different organizations and fields and the active participation of private-field businesses.

When data are increasingly shared and linked across different fields, it will lead to the formation of a data ecosystem (an economic ecosystem mediated largely by data). Such an ecosystem consists of data providers and data users. The data users can be businesses that deliver services to residents (such as transport services or healthcare). Data sharing can enable these users to improve the services they deliver or to reduce the costs they bear in delivering the services. The fundamental community functions and services that enable this ecosystem to work are data infrastructures. Data infrastructure includes the networks through which data are sent, the platforms on which data are stored, and the marketplaces where data are traded. This chapter discusses data infrastructure and the data ecosystem and their effective application in smart cities.

In addition to discussing how data can be used to generate value, the chapter also discusses personal data protection and other ethical matters related to the use of data. These ethical matters are also discussed in Chap. 6—the chapter on data governance (one of the six key factors).

10.2   What Is a Data Ecosystem?

A data ecosystem is a set of social connections through which companies and other organizations share data with each other and through which valuable services are delivered and the value of such is consumed. This section clarifies how data are exchanged and used to give readers an understanding of a data ecosystem.

We shall start with the types of data that are sources of value. Several different types of data exist, but the types that are exchanged include industrial data (data organizations use in their business operations) and personal data. In some cases, industrial data was once personal data. The types of data are described below.

Industrial data: This is data generated in the course of work and activities conducted by an organization. Business operators delivering services related to social infrastructure (utilities, transport companies, communications services, and so on) generate and store data about their equipment and operations, along with data about the use of their services. Industrial data can also include local government data, such as local topographic or climatic data and demographic or socioeconomic statistics.

Personal data and databanks: Some data are personal data, namely, data about private citizens (individuals and households). The challenge with personal data is that the person's privacy and interests must be safeguarded before it is collected and used. An example of personal data is vital data obtained from sensors that record the person's physical activity. Such data has commercial value; it can be used by healthcare providers for a start. However, while business operators such as drug companies might be interested in such data, they can scarcely collect the data by themselves. The process of collecting personal data from a large sample of people is challenging by itself; on top of that, numerous precautions, ethical considerations, and processing are needed to be undertaken to safeguard the privacy and rights of the individuals concerned. Companies will therefore entrust all or part of this work to a databank (Ministry of Internal Affairs and Communications 2022). Databanks hold the data of individuals and organizations, allow use of the data, and then return the profits from such use to the owners of the data.

In a smart city, such data will be used to generate new value. What does this mean exactly? It means that the use of the data is not retained by the original organization or individual. It means that the use of the data is extended, with the consent of the parties concerned, to organizations and purposes other than the original one. An example is when a supermarket chain or restaurant business decides on when and where to open new locations based on data that mobile carriers sell, namely, origin-destination flow data derived from the GPS data in people's smartphones.

How can the use of such data contribute to the development of local communities and society as a whole?

Figure 10.1 presents an example of how a data ecosystem can develop around AI-driven cameras as a public amenity. In this scenario, CCTV cameras have been installed as an anti-crime measure in several locations in an area with the consent of citizens and other stakeholders. The cameras provide safety and security, as well as help prevent crime. They alert security staff when they detect an abnormality, such as someone crouching down. What else can the data be used for? The camera footage could be analyzed to create data about the flow trends among passersby. If flow trends are measured continuously, it would be possible to observe flow patterns specific to certain days of the week, certain times of the day, or certain weather conditions. This information would prove useful marketing data for local restaurants and retailers (Municipality of Copenhagen and Capital Region of Denmark 2018). Using predicted volume of crowd, a restaurant might avoid purchasing excess foodstuffs. With data showing when a surge in demand is likely to occur, a business will be better able to gear up for the demand spike and capitalize on the revenue opportunity. Alongside cameras, LiDAR (a technology that emits a laser beam to determine ranges and positions of objects) could be used to create a map of people flows in the target zone and discover the locations where people tend to stop;

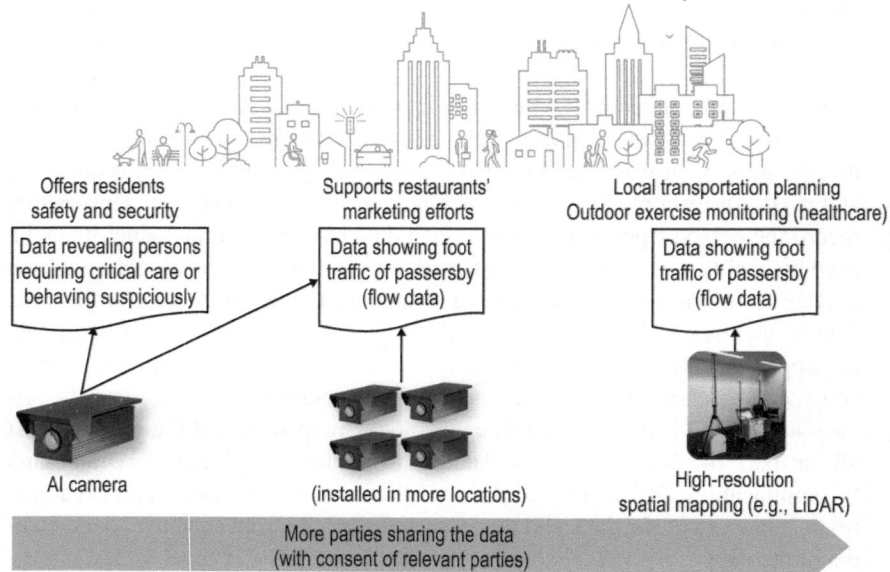

Fig. 10.1 Example of expanded use of data (AI camera scenario)

this information could serve as basic reference data for effective advertising and event planning. By organizing community events and providing amenities and infrastructure (such as benches) in locations where people tend to stop, we can attract people who enjoy spending time in a comfortable, safe environment.

Unlike in the case of physical objects, data can be replicated with minimal cost, meaning that a piece of data can generate fresh value repeatedly. What facilitates the use of such data in the community as a whole is data infrastructure. Data infrastructure encompasses various functions and measures; building a full set of data infrastructure is not a good way to facilitate data use. It is necessary to start with a small ecosystem and then gradually build up the infrastructure in tandem with the growth of the ecosystem. The measures for doing so are discussed below.

10.3   Social Infrastructure Supporting Data Use: Data Infrastructure

In the previous section, we mentioned that data infrastructure supports the use of data. Infrastructure can include the services that underlie social activities, including utilities (electricity, gas, and water) and transport services. Data infrastructure, though, is rather abstract and hard to visualize. To help illustrate the nature of data infrastructure, Fig. 10.2 compares data infrastructure with electricity infrastructure.

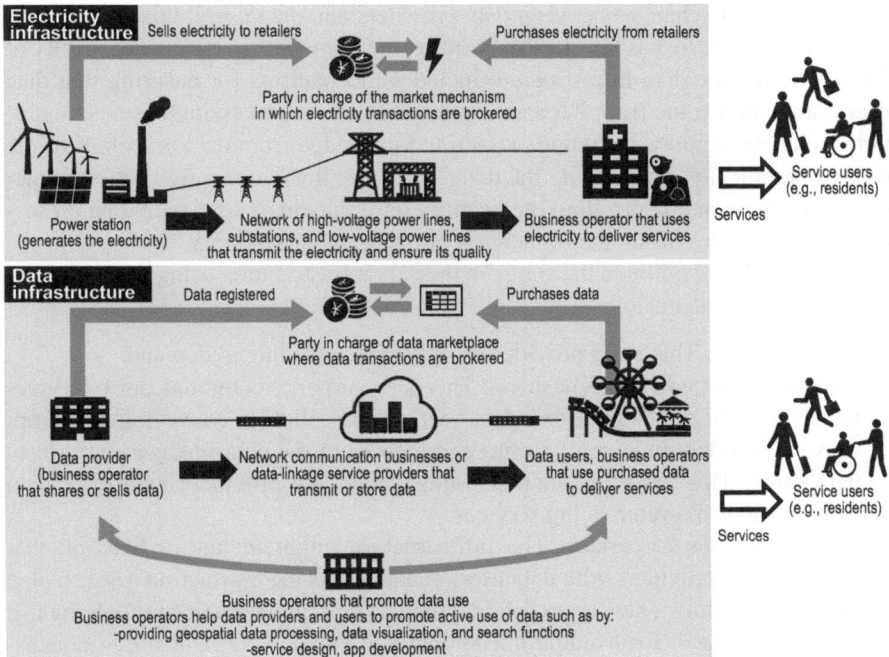

Fig. 10.2 Overview of electricity infrastructure and data infrastructure

What do you imagine the electricity infrastructure consists of? It consists of the following set of processes: Once electric power is generated in a power plant, it is transmitted across power lines over a long distance. The high-voltage electricity is then transformed to a lower voltage in a substation and then transmitted through a local network of power lines (those that are common features of villages, towns, and cities) to the demand side (residential and commercial buildings). To understand data infrastructure, one must also understand the contractual aspects underlying the infrastructure. Herein operates the market mechanism that brokers electricity transactions and balances power supply with power consumption. Specifically, electricity retailers anticipate the total amount of electricity that will be consumed by the numerous electricity consumers they have a supply contract with and then purchase an amount of electricity to accommodate this demand. The electricity is purchased from electricity-generating businesses. Market mechanisms determine the target volume of electricity generation. The infrastructure also includes functions that guarantee the quality of the electricity (its voltage, current frequency, and waveform) along with safety functions (rules, circuit breakers).

Infrastructure supporting data use has a similar pattern; like electricity, data flow from data providers to data users, with the data in this case transmitted by network communications businesses and data centers. Similar to the electricity market, there is a market for exchanges between data providers and data users; this is called the data marketplace. Just as the electricity infrastructure works to ensure the quality of the electricity, the data infrastructure includes mechanisms for ensuring that data users can obtain value from the data. For example, data processing businesses provide technical support, including extracting knowledge from data on behalf of the user or graphically representing the data. To ensure the data are used safely, functions governing and controlling the proper use of data exist (personal data protections, for example) inside the framework of operating a smart city.

Below, we have outlined the actors in the data infrastructure, using the electricity infrastructure as an analogy.

- Data provider: This party provides or sells data according to demand.
- Network communications business: This party provides communication services across a range of channels (wireless, wired, and cellular), conveying data from sensors to a data-linkage platform, and then from the data-linkage platform to data users. This party includes data-linkage service providers who offer data storage and cross-referencing services.
- Data marketplace operator: The infrastructure might include a function that matches data providers with data users and also sets the transaction price. It also assists in creating a new approach to demand–supply matching (through the use of a needs board, for example). The data marketplace operator might concurrently operate as a data-linkage business.

Data use is currently at a stage of searching for opportunities to use data and methods for using it. The key lies in matching data providers with data users. However, whereas in the case of electricity, one can walk into a consumer electronics store and pick a range of devices to utilize electricity, in the case of data, the ways of using data remain underdeveloped, so some extra work is required to match providers with users.

Several measures exist to achieve this. we have listed some of them below, along with the parties responsible for implementing the measure.

- Data-use platform operator: This party provides communication functions such as geospatial data processing, data visualization, and search functions.
- Service-design consultants, app-development consultants: These parties extend the range of opportunities to use data and design systems and services for service providers.
- Analytics provider: This party provides the analytics services required for using data.
- Databank: This party assists in the use of personal data requiring careful handling. A databank holds an individual's data and allows a third party to use the data within the permitted scope, ensuring that the data are always used in accordance with the person's consent (which may include specifying the purpose for using the data).

- Area management organization (in the case that the data infrastructure is operating in a redevelopment zone): This party designs and runs measures to promote data use and provides governance for appropriate data use (through ethics committees and other mechanisms).

Because data infrastructure remains in a stage of development, the term typically connotes the data marketplace and other technical platforms and services linking data providers and data users. In this chapter, we use the term in a broad sense, encompassing data providers, networks, data centers, data-use platforms, service design, app development, analytics, databanks, and data governance.

Because it is not possible to establish the whole data infrastructure in one go, it is necessary to start by putting in place the central roles and then gradually expanding the functions in tandem with the growth in data use. What, then, are the central roles? They are explained in the next section. Before that, the following article summarizes the relationship between data infrastructure and an urban operating system.

Column: Relationship Between Data Infrastructure and an Urban Operating System

How does such data infrastructure relate to an urban operating system? (Cabinet Office 2023). The relationship is illustrated in Fig. 10.3. An urban operating system's core functions include data management and external data linkage (part of data linkage). These functions overlap with the data-linkage services in the data infrastructure explained earlier; they are the principal functions in data infrastructure. As long as these functions exist, data is technically interoperable between all the different organizational actors. An urban operating system has, in addition to these functions, functions such as service linkage, authentication, and service management. Thus, some of the urban operating system's functions are also components of data infrastructure.

On the other hand, data infrastructure also includes some functions that are not generally part of an urban operating system. These include the function of matching data providers with data users and setting prices (an example being the data marketplace), as well as the function that offers a new approach to linking data (an example being a needs board that lists the types of data users want to use). They also include the data-provider system. As we deem the data-provider's system part of the data infrastructure and not part of the urban operating system, it is (from the perspective of the operating system) an external system.

The data infrastructure functions (or the functions that help grow the data ecosystem) introduced in Sect. 10.4 must be implemented and operated separately from an urban operating system. However, insofar as the urban operating system is a service platform, the functions promoting data use can also operate as services of the urban operating system.

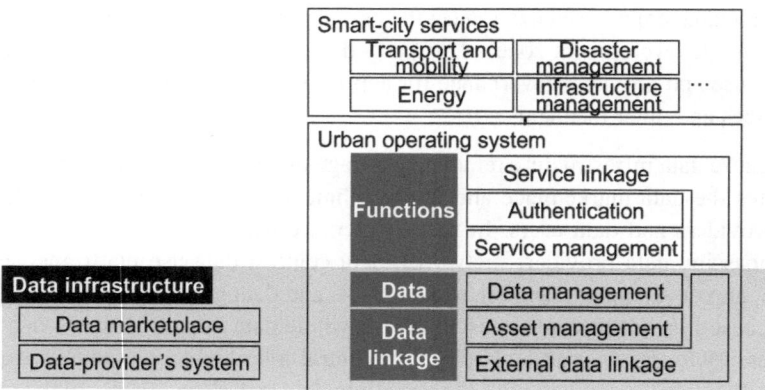

Fig. 10.3 Relationship between data infrastructure and an urban operating system

10.4 Data-Flow Functions and Measures for Promoting Data Use

While one may scarcely notice it in our everyday lives, numerous platform services for using data have been piloted and commercialized in Japan and around the world. Many of these services are designed to promote both data provision and data use. The services include search engines and other data retrieval systems. They also include community activities that bring together relevant parties.

In our Habitat Innovation project, we analyzed public-domain examples of data-use initiatives in Japan and overseas in order to gain insights into measures for promoting data use. We identified 20 types of measures, which are listed in Fig. 10.4. Some measures involve community-focused activities for enlisting the key stakeholders. We grouped these activities under the category "anthropological measures." Some measures involve IT systems and services. We grouped these activities under the categories "functional measures" and "interface measures." We shall outline each category in turn.

Anthropological measures are particularly important when the data ecosystem is in its early stages following launch. The key during this stage is to devise and implement scenarios for data use among the still-small set of data providers and users. Hence, you would need to use anthropological measures such as organizing events to motivate the provision and use of data, providing co-creation spaces and communities that use specific types of data, running a citizen-participation program, and helping volunteers address needs. These measures can be augmented by deploying the talent who can prepare open data that allow for a wide range of uses and by incentivizing companies outside the community to participate in the ecosystem; such efforts will prove effective in bolstering the above measures.

Other anthropological measures include publicly announcing the data owners' data and clarifying in writing the rights of and benefits for citizens; such measures

Category	Approach	Program	Example
Anthro-pological measures	Communicating with society	Publicly announcing data disclosure	[1]
		Clarifying, in writing, the rights of and benefits for citizens	[1]
	Talent deployment	Deploying the talent for open data	[1]
		Incentivizing the participation of companies outside community	[2]
	Talent development	Program for training developers	[2]
		Events to motivate the provision and use of data	[3]
	Community building	Expanding range of user-oriented functions	[3]
		Providing co-creation spaces and communities that use specific types of data	[4][5]
	Addressing social needs	Running a citizen-participation program	[1]
		Helping volunteers address needs	[6]
		Exploring use cases	[3][7]
Functional measures	Providing complex data with value	Logging tailor-made data in data catalog	[8]
		Providing samples of anonymized citizen data	[9]
	Data providers	Framework for encouraging individuals to provide personal data requiring protections and considerations (e.g., the Act on the Protection of Personal Information)	[10]
		Design the incentive system (monetary compensation, points)	[10]
		Effective support for pricing (with reference to other transaction prices)	[3]
	Strengthening stakeholder relationships	Consolidation of data on local needs	[6]
		Logging and matching users of data catalog	[2]
	Enhancing data linkage	Data users improve metadata (data about the data)	[8]
		Collating datasets to obtain insights for obtaining high-quality data	[11]
Interface measures	Data access/ discovery	Cataloging, search engine	[1][3]
		Plotting onto a map	[1][3]
		Providing sample data	[10]
		Providing data jackets (a summary of the dataset) that highlight the advantages of the data	[11]
	Analytics	Supporting linkages between applications	[1]
	Data transactions	Simplifying the data transaction process	[12]

Fig. 10.4 Summary of measures for promoting data

Example	Reference
[1] NYC Open Data	(City of New York 2024)
[2] Hiroshima Sandbox	(Bureau of Commerce, Industry and Labor, Hiroshima Prefectual Government 2024)
[3] EverySense	(EverySense 2023a)
[4] Grid Data Bank Lab.	(Grid Data Bank Lab. 2024)
[5] Common Platform, Saitama Edn	(UDCMi 2022)
[6] Data Science for Social Good Initiative	(Data Science for Social Good Initiative 2024)
[7] Okinawa Data Platform	(ISCO 2021)
[8] EverySense, SoftBank Corp., NS Solutions	(EverySense, SoftBank Corp., NS Solutions 2020)
[9] Hack My Tsukuba	(Tsukuba City 2023)
[10] EverySense	(EverySense 2023b)
[11] Innovators Marketplace on Data Jackets	(Innovators Marketplace on Data Jackets: The latform for solving social problems by data cooperation 2024)
[12] Data Marketplace Narrative	(Data Marketplace Narrative 2024)

Fig. 10.4 (continued)

will promote greater use of data, raise awareness about participants' contributions to the community, and extend the range of activities.

Next are functional activities. What functions of an IT system can complement anthropological measures? Several system functions can play a role, not so much in supporting the process of devising ideas for using data but more in supporting processes that take place after an idea is pitched. For example, once an idea is pitched, sample data can be used to show exactly how the idea could be implemented; once the idea proliferates, a search engine can be used to help people find the right data from among a vast array of data.

One effective measure, for example, is to provide samples of anonymized citizen data to promote the development of use cases. In this way, we could, before the data are received, glean from the data format, data frequency, and density the extent to which analytics and visualization are possible. Once the ecosystem has grown in scale to some extent, the following measures will prove useful for developing effective use cases: consolidation of data on local needs, logging and matching users of the data catalog, data users improving metadata (data about the data), and collating datasets to yield insights for obtaining high-quality data.

Once usage proliferates, measures to extend the range of the activities should be undertaken, such as providing a framework for encouraging individuals to provide personal data requiring protections, incentivizing data provision (monetary compensation, points), and providing effective pricing support. These measures will be demanded by many users.

Finally, there are interface measures, which are designed to ensure the effective use of data, especially when there is a large array and volume of data. These measures include providing cataloging and search engine functions covering all the available data, functions that plot data onto a map, providing sample data, and providing data jackets (a summary of the dataset) that highlight the advantages of the data.

It would be neither efficient nor effective to launch all these anthropological, functional, and interface measures for facilitating the ecosystem in one go. It is far better to implement them gradually, in tandem with the growth of the ecosystem. During the ecosystem launch phase, it is important to fully verify each data-use case one by one to ensure that they can be rolled out effectively. This entails prioritizing anthropological measures that generate data-use cases. Next, once the range of data providers and data users has increased, improving the usability of the available data and providing ongoing procedural support related to the use of the data should be the focus.

10.5   Data Provider and Users' Needs and Impediments

In the previous chapter, we discussed measures for facilitating the growth of a data ecosystem. In that discussion, we had envisaged a generic, hypothetical local community. When it comes to the use of data in an actual community, what challenges and impediments will arise in growing the ecosystem? In our Habitat Innovation project, we asked Kashiwa-no-ha Smart City Consortium about the motives, the purposes, and the impediments to using data. We shall discuss the results of the study in this section.

We conducted the study in 2022. We asked Kashiwa-no-ha Smart City Consortium and related organizations questions in an online survey and obtained 25 responses.

The first question concerned motivation; respondents were asked to describe what they, as a data provider or data user, hoped to achieve by using data. Most respondents were business operators, so we assumed that they would see a data ecosystem as an opportunity to grow their business or make a profit from data. However, the responses confounded these expectations (see Fig. 10.5). The top motivation for using smart-city data was to contribute to the community (such as to help the community thrive or address the needs of citizens). The second biggest motivation was to collaborate and innovate with other organizations. Far fewer respondents, by comparison, cited a motive directly related to making a profit for their own organization (making a profit from data or capitalizing on an opportunity for business growth).

To gain further insights into the strength of motivation, the next two questions asked respondents to rate their degree of interest on a 5-point scale. The first question asked respondents to rate their interest in data accumulated in any given smart city (the question did not specify any particular smart city). We determined that the level of interest in using data was generally high. The second question referred to a

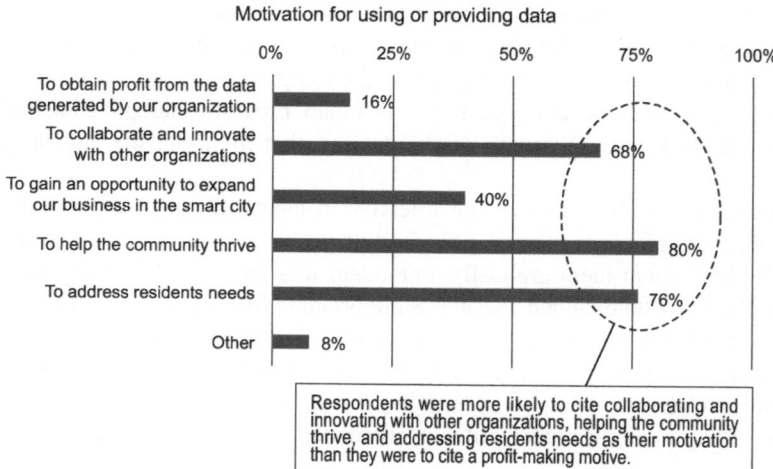

Fig. 10.5 Distribution of responses about key motivations for providing or using smart-city data (multi-choice question format, $N = 25$)

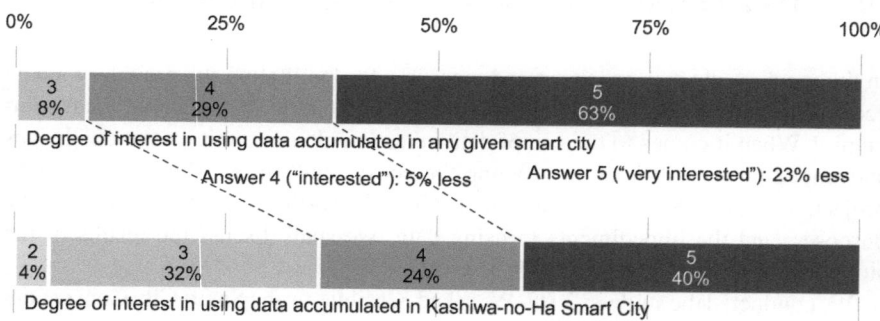

Fig. 10.6 Variation in degree of interest in using data: comparison between responses for smart cities in general (top) and those for a specific project (bottom)

specific smart city; it asked respondents to rate their interest in using data specifically in Kashiwa-no-ha Smart City. The results indicated a slightly lower degree of interest compared to that obtained for the previous question (Fig. 10.6). These results imply that while businesses are interested in using data in general, when it comes to using data in a specific community, problems in generating and implementing ideas for specific data use may arise.

The next question asked data users about the impediments they face in using data. The responses are shown in Fig. 10.7. Respondents were most likely to say that they had no idea of how to use the data. Other answer options with a large number of responses were "We are worried about whether we would have proper authorization to use the data or whether we could safeguard personal data properly" and

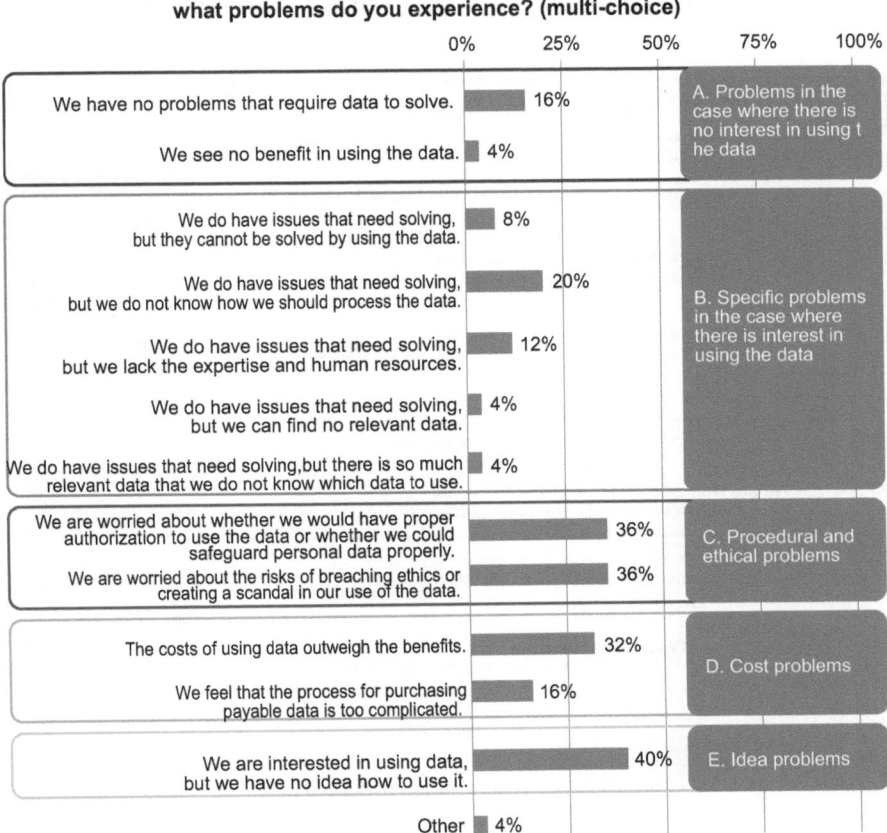

Fig. 10.7 Distribution of responses regarding issues experienced in the use of data

"We are worried about the risks of breaching ethics or creating a scandal in our use of the data." Another popular response was that the costs outweigh the benefits.

The procedural and ethical problems that many of the respondents cited comprised two issues: the need to ensure the user has the proper authorization and necessary safeguards for personal data, and the risk of ethical breaches or scandals. The high level of concern we observed for these issues is closely related to two other key factors: social acceptance and data governance.

The final question asked data providers about the problems that impede them from providing data. The responses are shown in Fig. 10.8. The most commonly cited problem was concern over whether the user would have proper authorization to use the data and whether it could safeguard personal data properly. Thus, helping businesses address this issue could encourage the use of data in the community and its provision.

The next most commonly cited response was "we are unsure of what data to provide."

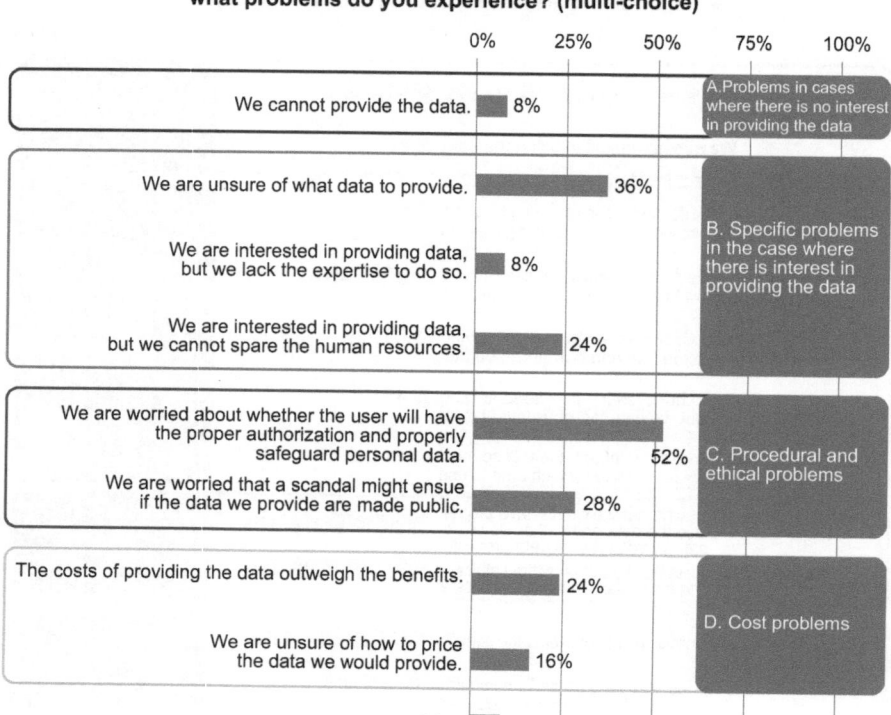

Fig. 10.8 Distribution of responses regarding issues experienced in the use of data

From the results of our study, we derived two measures for encouraging the use of data.

10.5.1 Measure 1: Support the Generation of Ideas for Using Data to Address Community Needs

The purpose of this measure is to help stakeholders generate ideas for using the data to address community needs. The study revealed that while some 70–80% of respondents wanted to use the data to contribute to the community (to help it thrive or to address the needs of citizens), the most commonly cited impediment to using the data was the lack of ideas on how to use it. To help generate new ways to use data, we suggest presenting specific needs from the perspective of citizens, considering how these needs can be addressed, and then linking this discussion with a showcase of examples of data use relevant to the attributes of the data in question. It would also be effective to hire a consultant to assist in the process of thinking about the data and ideas for using it, or to use a living lab (see Chap. 7 for more on living labs).

10.5.2   Measure 2: Support from the Party That Wants to Use Data

The purpose of this measure is to encourage the party that wants to use the data (as opposed to the party that wants to provide it) to communicate their demands to data providers in order to help create more data-use cases. The second most commonly cited problem among data providers was that they were unsure of what data to provide. We also found that organizations tended to show greater interest in using the data than in providing it. Therefore, when the party seeking to use the data communicates their specific requirements to solve a particular problem, data holders will have a better idea of the value of the data they hold, contributing to a mindset that is more willing to provide data.

Another potentially effective measure is one that relates to data governance, which is one of the key factors. The measure is to create a framework in which people can easily check whether the data have the necessary authorizations and privacy safeguards. Both data users and providers expressed concerns about processes and ethics. These fears can be addressed by providing an approachable helpline whereby parties who want to use data and those that want to provide it can seek expert advice about these concerns. Alongside this, providing an FAQ page might help lower the hurdle for newcomers. These measures relate to two other key factors: social acceptance (see Chap. 5) and data governance (see Chap. 6).

In Kashiwa-no-ha Smart City, AI cameras are used to analyze footage and monitor foot-traffic trends at a number of locations in the area (UDCK Town Management 2021). In our Habitat Innovation project, we have launched two programs to encourage the use of this data. The first involves co-creating services with citizens in a living lab. The second involves using data-visualization tools for prototyping tools and piloting measures for encouraging their use.

In the first program, the participants pitched ideas about the kind of applications they would like to see and the kind of data they would like to obtain from the AI cameras. However, since no such data existed at the time of the workshop, it was hard to advance to a discussion of how the ideas could be turned into feasible services. We therefore prepared sample data for the ideas (for example, we prepared a sample of data about dog walkers, one of the ideas being about a service for dog walkers) and used tools to visually represent the data. Of the measures for encouraging data use, this program corresponded to the measure of providing sample data. In this program, we reaffirmed how important providing sample data is in generating ideas for using data, and we also learned about the importance of tools for efficiently generating sample data.

10.6 For Continuous Value Creation

For the final part of this chapter, we shall continue the discussion about the outlook for the AI camera initiative mentioned at the start. To recap, the cameras were initially installed to safeguard the security of residents. However, their use was later expanded after more cameras were installed and LiDAR was deployed to enable precise monitoring of foot traffic. The cameras were now being used not just for security but also for informing efforts to facilitate production of interesting events and amenities where they were needed, thereby enhancing comfort and safety of the community. Figure 10.9 shows a value loop in which data are used with the consent of citizens to create services that deliver value back to them. The use of foot-traffic data does not need to stop there. Its usage could be extended into the traffic and healthcare fields; for example, it could also be used to predict traffic congestion hotspots so that the congestion can be eased or for generating insights about what health management services could encourage people to walk about comfortably. Conceptualizing such foot-traffic data as a seedling in an ecosystem, we discussed how the ecosystem can be grown.

Growing the ecosystem requires more than simply laying down the infrastructure; it necessitates choosing measures for promoting data use and implementing the measures effectively. Specifically, in the early stages, focusing on platform functions and analytical functions is less of a priority than devising, implementing, and evaluating use. While the effects may be limited, implementing use cases is

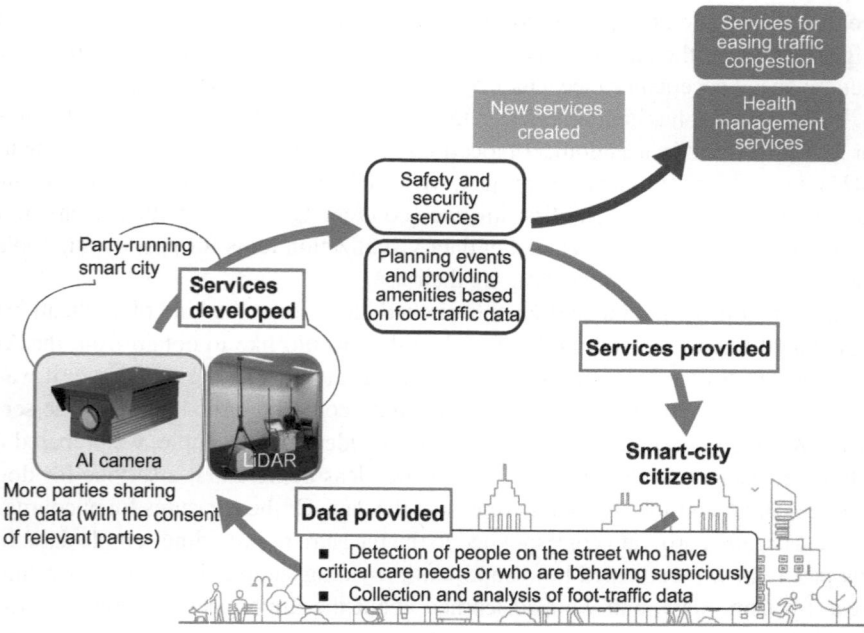

Fig. 10.9 Vision of a data ecosystem that develops from AI cameras

relatively straightforward and gives stakeholders a chance to see for themselves how the data can be used; from there on, the process can gain traction.

Then, as the range of data-use cases expands, attention may be directed on leveraging infrastructure functions to mass produce some of these data-use cases. The task of exploring data-use cases for this purpose should not be undertaken just by the consortium members or the target community; it is important to solicit ideas from a wide range of parties.

When we talk of continuous value creation, what does "continuity" mean? Does it mean that a service that seems to be working effectively should be delivered in perpetuity? If a service is benefiting citizens, it should obviously be continued, but there is no guarantee that a particular need will remain forever. Given that needs change over time, continuity must mean adapting to the changing needs with the right kind of metabolism in services. Leaving aside the matter of whether citizens are aware of the needs in question, it is the citizens, as the real beneficiaries of the data, who judge whether or not a service is valuable. It is therefore necessary to engage with the citizens when evaluating or refining existing services, when introducing new programs, and—when necessary—replacing services.

The process of continuously creating value is iterative: Once the outlook becomes clearer, the first step can be taken—something that can be done with just a small team. The benefits of this first step can be shared with both the service providers and the citizens, and then the activities can be expanded slowly but surely.

10.7 From Key Factors to Applications

In Part II (Chaps. 4, 5, 6, 7, 8, 9, and 10), we discussed the six key factors (social acceptance to the data ecosystem) with reference to the research findings of H-UTokyo Lab. Here, it is worth recapping the two main ways in which the six factors are important.

First, the six factors complement the Society 5.0 reference architecture by presenting a construction method and delivery roadmap. Second, they are necessary for building a people-centric, sustainable smart city.

In Chap. 2, we discussed how, with smart-city projects underway in Japan and around the world, we are in the midst of a trend whereby digital technology and data are being used to transform cities. we also argued that the key factors have an important role to play in ensuring that these smart-city projects, to which so much financial resources and labor are being devoted, do not end up a passing fad.

Across Japan, we witness repeated efforts to promote smart-city initiatives, with national subsidies being provided for rollout or experimental projects. The focus here tends to be on ensuring the continuity of the business or service itself. However, the critical question for any smart city is how these undertakings can be sustained, either as integrated smart-city services or as an accumulated set of services that operate independently of each other. The six factors need to be applied effectively to pave the way toward a sustainable smart city.

However, the key factors are not projects in and of themselves; they are things that accompany the actual efforts to develop smart-city businesses and services. The challenge, then, in a project to create a people-centric, sustainable smart city is to work out the delivery of the project in a way that addresses actual community needs and how the Society 5.0 principles can be applied to these needs. Therefore, there is a need to see some application cases illustrating how the key factors should be applied.

To that end, Part III (Chaps. 11, 12, and 13) presents three projects ("public dialog in data-driven urban planning," "smart aging," and "value-creating infrastructure management") to illustrate smart-city themes that are common across Japan and how the key factors can be applied to each of these themes in a way that aligns with the principles of Society 5.0. Each of the projects was led by the H-UTokyo Lab. We hope they will serve as useful application cases, guiding efforts to apply data to an existing problem in the target city or community in a way that seamlessly integrates cyber and physical spaces—or, to use the phraseology of the 5th Science and Technology Basic Plan, addressing the problems with "the high degree of merging between cyberspace and physical space."

References

Bureau of Commerce, Industry and Labor, Hiroshima Prefectual Government (2024) Hiroshima Sandbox. https://hiroshima-sandbox.jp/. Accessed April 18, 2024.

Cabinet Office (2023) *Sumāto Shiti: Rifarensu ākitekucha howaito pēpā* [Smart city: Reference architecture white paper], 2nd edn. In: Cross-ministerial Strategic Innovation Promotion Program (SIP) Second Phase, Big-data and AI-enabled Cyberspace Technologies/Smart City Architecture Development/Smart City Architecture Design and Promotion of Related Verification Research (Released on August 10, 2023). https://www8.cao.go.jp/cstp/stmain/20230810smartcity.html. Accessed March 7, 2024.

City of New York (2024) NYC Open Data. https://opendata.cityofnewyork.us/. Accessed April 18, 2024.

Data Marketplace Narrative (2024) https://www.narrative.io/data-marketplace. Accessed April 19, 2024.

Data Science for Social Good Initiative (2024) https://www.datascienceforsocialgood.org/. Accessed April 18, 2024.

EverySense, SoftBank Corp., NS Solutions (2020) *Sumātoshiti ni okeru pāsonaru dēta to sangyō dēta no dēta torihiki shijō ni yoru kyōyū kiban no jisshō* [Verification of a common platform with a data trading platform for personal data and industrial data in smart cities]. In: *Senryakuteki Inobēshon Sōzō Puroguramu (SIP) dai 2 ki: Biggudēta AI o katsuyōshita saibā kūkan kiban gijutsu/sumātoshiti jisshō kenkyū* [Cross-ministerial Strategic Innovation Promotion Program 2: Verification-unnresearch for Big Data- and AI-powered Cyberspace Infrastructure and Smart Cities]. March 2020. https://www8.cao.go.jp/cstp/stmain/a-2-8_200318.pdf. Accessed April 19, 2024.

EverySense (2023a) *Kigyō kan chikuseki-gata dēta torihiki shijō EverySense Pro* [EverySense Pro: A Data Marketplace with Inter-company Clustering]. https://every-sense.com/products-services/everysense_pro/. Accessed April 18, 2024.

EverySense (2023b) *IoT dēta torihiki shijō EverySense* [EverySense: An IoT Data Marketplace]. https://every-sense.com/services/everysense/. Accessed April 18, 2024.

Grid Data Bank Lab. (2024) https://www.soumu.go.jp/main_content/000683154.pdf. Accessed April 18, 2024.

Innovators Marketplace on Data Jackets: The platform for solving social problems by data cooperation (2024) https://imdj.datajacket.org/. Accessed April 18, 2024.

IT Innovation and Measure Center Okinawa (ISCO) (2021) Okinawa Data Platform. https://isc-okinawa.org/blog/odpfseminar-20210128/. Accessed April 18, 2024.

Ministry of Internal Affairs and Communications, Regional Communications Development Division (digital corporate behavior office) (2022) Jōhō ginkō no torikumi [Databank Initiatives]. January 2022. https://www.soumu.go.jp/main_content/000791752.pdf. Accessed April 18, 2024.

Municipality of Copenhagen and Capital Region of Denmark (2018) City Data Exchange–Lessons Learned from a Public/private Data Collaboration. March 2018 https://cphsolutionslab.dk/media/site/1837671186-1601734920/city-data-exchange-cde-lessons-learned-from-a-public-private-data-collaboration.pdf. Accessed April 18, 2024.

Tsukuba City (2023) *Hack My Tsukuba: Kadai kaiketsu-gata wākushoppu* [Hack My Tsukuba: A Problem-Solving Workshop]. https://www.city.tsukuba.lg.jp/shisei/joho/1008026/1008220/1008123.html. Accessed April 18, 2024.

UDCK Town Management (2021) *AI kamera o katsuyōshita "anshin anzen" na machi no mimamori sābisu* [Using AI Cameras for a Neighborhood Watch Service for a Safe and Secure Community]. https://www.udcktm.or.jp/ai/index.html. Accessed April 18, 2024.

Urban Design Center at Misono (UDCMi) (2022) *Kyōtsū purattofōmu Saitama ban* [Common Platform, Saitama Edn]. https://www.misono-tm.org/udcmi/projects/61.html. Accessed April 18, 2024.

Part III
Applications for Smart Cities Toward Society 5.0: H-UTokyo Lab's Initiatives

Chapter 11
Project 1: Public Dialog in Data-Driven Urban Planning

Shin Osaki and Yuki Igeta

Abstract This chapter contains two case studies around Public Dialog in Data-Driven Urban Planning.

The first case is about data visualization. The visualization and the dialog around it inspired ideas and feedback from local stakeholders. This suggests that a data-driven city is not just led by smart city operators but also by the community with appropriate data representations. We have seen the rise of a kind of community-level, grassroots leadership, which we should perhaps dub data-driven local management.

The second case is about hybrid (online and offline) approach to the public dialog. The hybrid approach can throw open the doors to people who were unable to participate in conventional in-person workshops. Our workshops were attended by graduate students, housewives, and other people who were participating in community building for the first time, via social media. Hybrid dialog increases the range of participation and thus the range of perspectives represented.

From our observations about data visualization and public dialog, we concluded that combining the two into a single platform will help diversify the range of people participating in data-driven local management. Under these conditions, public dialog will go beyond the task of simply gaining public consent toward a proposed intervention and become a step toward a more creative and more grassroots form of community building.

Keywords Data-driven urban planning · Data-driven local management · Public space · Hybrid public dialog · Digital communication platform

S. Osaki (✉)
Neighverse Inc., Tokyo, Japan

Sustainable Society Design Center, Graduate School of Frontier Sciences, The University of Tokyo, Tokyo, Japan
e-mail: osaki@edu.k.u-tokyo.ac.jp

Y. Igeta
Graduate School of Frontier Sciences, The University of Tokyo, Tokyo, Japan
e-mail: igeta.yuki@edu.k.u-tokyo.ac.jp

Hitachi-UTokyo Laboratory (H-UTokyo Lab.), *The Architecture of "Society 5.0"*, https://doi.org/10.1007/978-981-96-2929-9_11

161

11.1   An Attempt at Data-Driven Urban Planning During Covid

11.1.1   The Value of Public Dialog in Data-Driven Urban Planning

One type of smart-city initiative involves data-driven urban planning. In the process of planning new urban programs, one will use various kinds of urban data to make evidence-based decisions. The wide-ranging urban data includes foot-traffic data, vehicle-traffic data, and density data obtained from laser sensors or AI cameras; environmental data and energy data obtained from sensors installed in urban infrastructure; and purchase data obtained from electric payment services (with the user's consent). These datasets can be combined to simulate the possible outcomes of proposed urban programs, enabling judgments and decisions about these programs to be based on some degree of evidence. Until now, these simulations of future scenarios have been conducted through pilot programs. This approach enables the extensive preparation and testing processes to be streamlined and to base the tests on a range of hypotheses.

However, even if a huge array of data is cited to demonstrate the reasonableness of the intervention in question, as long as a top-down approach is taken, with decisions in closed-off settings, one is unlikely to gain public support. It is better to make the data public in a format that is visible and plain and to provide an interface for public engagement (opportunities for dialog) so that the program is advanced in a way that the public can support; this process is essential for ensuring social acceptance.

11.1.2   Two Covid-Era Projects to Encourage Public Dialog

Matsuyama City in Ehime Prefecture is working to become a data-driven city. Some years ago, Matsuyama City and Urban Design Center Matsuyama (UDCM) started using urban data to simulate programs. In the year that ended March 2018, we (H-UTokyo Lab) started developing a tool and method for facilitating dialog with the public.

We started engaging in the planning efforts in the year ending March 2021, when the Covid-19 crisis arose. The pandemic had forced Matsuyama City and UDCM to postpone data collection and various preparations in Matsuyama. They had hoped the project could be undertaken in outdoor settings to avoid the three Cs (closed spaces, crowded places, and close-contact settings), but the data users (parties involved in a shopping arcade) were struggling amid the lack of expert know-how. They then requested our assistance.

In view of the situation on the ground, we launched two projects simultaneously. The first project was designed to assist in the use of outdoor spaces where the three Cs (closed spaces, crowded places, and close-contact settings) could be avoided. We launched this project on the belief that we could encourage the use of outdoor settings by collecting outdoor foot-traffic data and then visualizing the data so that it could be used to guide discussions. The second project involved an online tool for facilitating public dialog. We launched this project on the belief that we could incorporate into public dialogs the advantages of online communication tools such as Zoom and other videoconferencing apps that had rapidly proliferated in the pandemic.

In the map of data-driven city planning (Fig. 11.1), the first project corresponds to data visualization, while the second corresponds to public dialog. These two processes should ideally be undertaken as part of a series, but because the pandemic had disrupted plans for on-site activity, they had to be undertaken as separate projects. This chapter introduces each project in turn, but please bear in mind that the projects should ultimately be incorporated into a single planning cycle for a data-driven urban city and undertaken in tandem with one another.

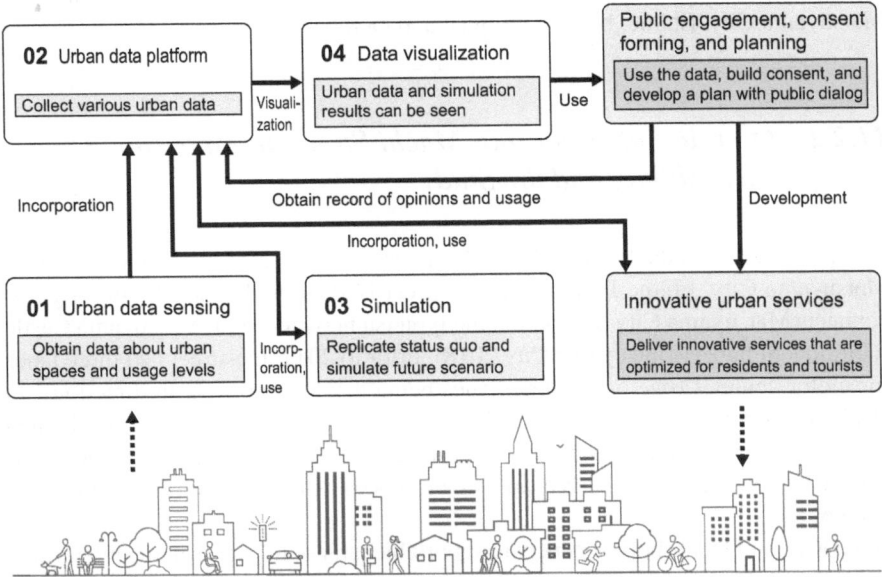

Fig. 11.1 Conceptual map for data-driven urban planning. (Ministry of Land, Infrastructure, Transport and Tourism 2019)

11.2   Visualizing Data Pertaining to Outdoor Public Spaces

11.2.1   Growing Interest in Outdoor Public Spaces: Background to the Project

Before the pandemic, outdoor public infrastructure (such as roads and public parks) had been garnering increasing attention for its role in creating a livable and thriving community. There were a number of models for using outdoor public spaces; New York (NYC DOT 2024) and Barcelona (Barcelona City Council 2024), for example, provided open cafes and organized art and music events in streets and public parks. Japan had also been arranging conditions for greater use of outdoor spaces by relaxing the criteria for having exclusive use of a street, for instance[1] (Ministry of Land, Infrastructure, Transport, and Tourism 2020). After the pandemic arose in 2020, the use of outdoor spaces received all the more attention globally amid the need to maintain economic and recreational activities while avoiding the three Cs.

Against this backdrop, we launched a project to promote the use of outdoor spaces as part of the data visualization theme in the cycle of data-driven urban planning. This project involved obtaining data on foot traffic in outdoor public spaces and data on the use of such spaces, visualizing such data, and then sharing it with the public and other stakeholders to gain insights into how outdoor public spaces can be used. Described below is a case that took place on Hanazono-machi Street in Matsuyama City, Ehime, between August and October 2022.

11.2.2   Overview of Hanazono-Machi Street (in Matsuyama City, Ehime) and the Study

The study was conducted on Hanazono-machi Street, a street in the center of Matsuyama City, Ehime Prefecture. The street is 40 m wide and 250 m long and connects Matsuyama City Station (on the Iyotetsu network) with Matsuyama Castle (Horinouchi Park) (Matsuyama City 2018). Prior to 2017, the street had three lanes for motor vehicles (one of the three lanes being a side road) in both directions. In 2017, the street was reopened with the three lanes in each direction consolidated into a single motor-vehicle lane in each direction, freeing up space for pedestrians and also creating a picnic area and other spaces where people can stop by or hang out.

UDCM and local shopkeepers placed portable tables and chairs in the picnic area and around the fixed benches so that anyone could access the amenities as they

---

[1] In June 2020, the Ministry of Land, Infrastructure, Transport, and Tourism relaxed the criteria for roadside restaurants and other premises to occupy the sidewalk outside to support businesses affected by the pandemic (Ministry of Land, Infrastructure, Transport, and Tourism 2020).

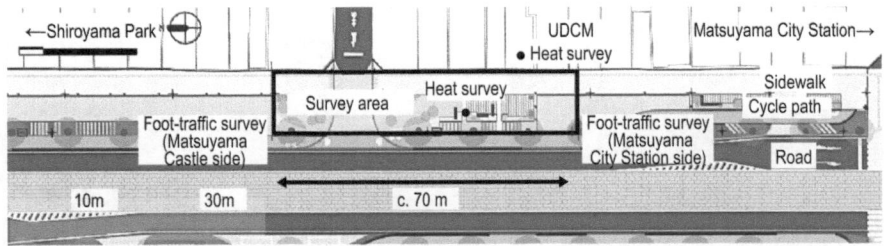

Fig. 11.2 Site of data collection (where the surveys were conducted)

desired. In September 2017, the revamped area held a marketplace event called the Ojoka Marche (Ojoka Marche website 2023). This was to be the first of many events run by local stakeholders: every Saturday, the area holds the Hanazono-machi farmer's market (Maipure (Matsuyama, Iyo, To'on, Matsumae, Tobe) 2023). Every fourth Sunday, the Matsuyama Hanazono Sunday market is held (Website for Matsuyama Hanazono Sunday Market 2024).

Our study focused on the eastern side of Hanazono-machi Street (the side where the picnic area and street furniture are located). We obtained data on how this space is used. Our data collection method was as follows.

We first used LiDAR to obtain data on people's behavior. LiDAR has minimal privacy issues because it does not record individuals' faces or voices as a video camera would; instead, it tracks people's movements by emitting a laser and timing how long it takes for the reflected light to return from the target. It is also able to map the movements of all individuals who enter the scanning range, meaning that, compared to GPS or Wi-Fi, its measurements will be much closer to the actual number of people. Moreover, compared to the method of having a member of the survey team use a manual foot-traffic counter, LiDAR can store continuous data showing people's precise movements. These advantages made Lidar suitable for obtaining insights about the use of outdoor public space.

Next, we obtained environmental data. These data were obtained in a heat survey in which a heatstroke meter was used to measure temperature, humidity, and a heatstroke index. We collected these data because temperature and similar environmental conditions significantly influence people's outdoor activities. Finally, to supplement the data, a questionnaire survey was administered to people using the site (Figs. 11.2 and 11.3).

11.2.3   Features of Public Spaces that Encourage People to Stop

To encourage outdoor activities, this project used data to demonstrate the environmental settings that increase the likelihood of people hanging out in Hanazono-machi Street. We focused on hang-out behavior because it can serve as a metric

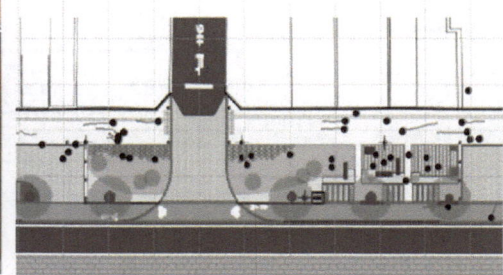

Fig. 11.3 Left: Location where the LiDAR mobile kit is installed, Right: LiDAR's 3D mapping of pedestrians

Default configuration   Uncovered furniture   Covered furniture

Fig. 11.4 Three configurations for the hang-out space (Left: Default configuration. Middle: Uncovered furniture. Right: Covered furniture)

indicating that the public outdoor area is being used as a place for dining, resting, or other forms of recreation. When we analyzed the data, we identified environmental features that encourage people to stop by or hang out in the space. Two of these features are described below.

First, we found seasonal variation in the public space configuration that will attract visitors. During the study, we changed the configuration each day, cycling between three different configurations (Fig. 11.4), and counted the number of visitors; a "visitor" was defined as a pedestrian who stays in the picnic area or bench area for at least 60 s. In August, the daily visitor total was highest on days when the space had covered furniture (tables and chairs covered with canopies). In October, this configuration had the opposite effect; visitor numbers were highest on days with the "uncovered furniture" (tables and chairs without canopies) configuration.

We observed similar results for duration by hourly time slot. In August, the duration of stay during 12:00–13:00 (when the midday sun was beating down) was ten times longer in the case where canopies were present compared to the other cases ("default" or "uncovered furniture"). In October, the duration was longer with the "uncovered furniture" (no canopies) configuration (Figs. 11.4 and 11.5). Predictably, canopies are preferred during the summer, while people prefer the absence of canopies in early autumn.

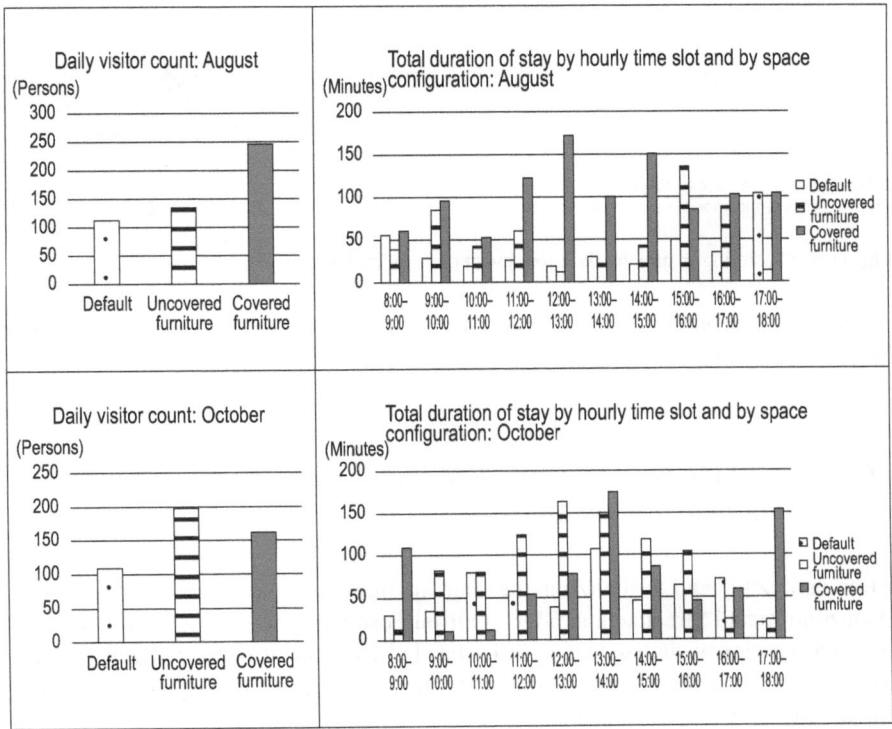

Fig. 11.5 Average visitor count and total visit duration in August and October 2022, by space configuration

Second, we observed considerable variation in visiting patterns (purpose and duration) between weekdays and event days. Visitor numbers were much higher on event days (Sundays) than they were on weekdays, but event days were associated with a shorter average duration of stay per person. On event days, we observed groups of people consuming takeout food purchased from a mobile food kiosk. Each group would soon leave, and another group would take its place. Thus, on event days, people were mainly looking for a place to consume refreshments, without needing to stay there for a long time. By contrast, on weekdays, we observed people staying for longer periods, working on a laptop, or waiting for someone. Thus, on weekdays, people were looking for a place to relax.

These trends remained consistent across August, September, and October. Thus, varying the spatial setup between event days and weekdays will make the space more accessible to users. For example, for event days, it would be a good idea to provide bendable benches. On weekdays, it would be a good idea to provide power outlets for laptop users.

In Fig. 11.6, we plotted durations of stay onto a map. The map reveals that duration of stay is associated with location of stay within the area: Short-stay visitors were distributed broadly across the whole area but had a relatively high

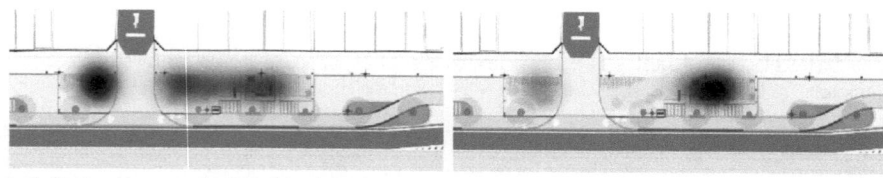

Left: Spots where people stayed for 1–3 minutes
Right: Spots where people stayed for longer than 3 minutes

Fig. 11.6 Clustering in area by duration: event days in September

concentration around the intersection. By contrast, long-stayers were concentrated in spots some distance away from the intersection.

11.2.4 Using Data Visuals to Facilitate Public Dialog About Outdoor Public Spaces

After analyzing and visualizing the data in this way, we began a dialog with Hanazono-machi Street stakeholders. Our purpose was to demonstrate the hypothesis that sharing visualized data with stakeholders can offer them fresh insights when they are discussing how outdoor public space can be better used. As anticipated, after the objective visual data was presented, a constructive discussion ensued, devoid of any digressions or rambling speeches. We also observed the potential for the data visuals to spark new ideas for discussion, including an idea for fresh data proposals. I shall introduce three examples of discussions on the use of outdoor public spaces with the owners of shops along Hanazono-machi Higashi Street and other stakeholders that ensued after the study.

Several of the stakeholders were already under the impression that, on weekdays, the hang-out area of Hanazono-machi Street is used not as a brief resting space but for medium and long durations (for work or as a place to wait for someone). When they were presented with data confirming that this was indeed the case, the stakeholders started discussing a branding/differentiation strategy for Hanazono-machi Street, including the idea of promoting Hanazono-machi Street as a place where people can hang out for longer periods compared to other shopping thoroughfares in the city.

The second discussion concerned the canopies covering the tables and chairs. The stakeholders had expected that providing canopies in summer months would encourage people to hang out in the area, and the data confirmed this assumption. After the data visuals were shared, the discussion shifted away from the matter of whether or not to provide canopies to how the canopies, if provided, should be used.

The third discussion concerned an idea for data linkage. Shopkeepers suggested that the data could be combined with sales data to identify uses of the space that could contribute to sales revenue. It was agreed that such analytics would provide a valuable resource for exploring ways to use the public space.

Thus, from our experiment with public dialog, we learned that presenting objective visuals that corroborate the stakeholders' subjective perceptions of a matter can contribute to a discussion in which everyone is on the same page, and that presenting visuals on patterns of hanging out in the area can inspire new hypotheses and new ideas for further data acquisition.

11.2.5   Data Visualization as a Community-Building Process

I shall end this section by discussing data visualization as a process for community building, including what we learned about organizing and running such visualization and the outlook for engaging in such visualization.

Data visualization involves obtaining and analyzing data and then presenting it to stakeholders to facilitate dialog. Insofar as data visualization is a community-building process, it should involve a continuous cycle of collecting, analyzing, and sharing the data; rather than being a one-off visualization for a single place and time, the cycle for several locations over several months or, if possible, years should, ideally, be repeated. Thus, the survey method would need to be made as efficient as possible and who is responsible for conducting it and who bears the costs should be considered. In the case of Matsuyama City, it was only after conducting a continuous survey for 3 months that we gained sufficient data for the visualizations and analysis that would stimulate an in-depth discussion among participants; having 3 months of continuous data meant we could account for seasonal and climatic factors as well as days of the week and intraday time. We identified issues concerning the organization and running of the data-visualization process: In this process, we relied on the support of many stakeholders; as our study made use of all the temporary apparatus, setting out the apparatus in the morning and clearing it away in the evening proved very onerous, time-consuming, and costly.

As for the outlook, data visualization is an effective tool for managing outdoor public spaces, and thus it will become increasingly important in the years ahead. In the case of Matsuyama, LiDAR measured foot-traffic data proved effective in analyzing use of the space. When foot-traffic data, instead of being used in isolation, is combined with environmental data (temperature, humidity), revenue data from the shops along the street, questionnaire survey data, and other datasets, it can lead to even more effective evaluation and enable more effective planning for a data-driven city.

Since the pandemic, people have placed more value in outdoor public spaces, and this trend is likely to continue. With the consent and support of residents, shopkeepers, and other stakeholders, data visualization can stimulate communication and encourage better use of outdoor spaces.

11.3   Hybrid (Offline–Online) Public Dialog

11.3.1   The Potential and Limitations of Online Dialog Tools

The pandemic prompted a dramatic rise in videoconferencing. Videoconferencing offers many benefits, such as allowing people to participate remotely in dialog with the touch of a button. However, some interactions will always be better in person. An in-person format is preferable when brainstorming ideas or discussing something in depth. We are also seeing the rise of privately organized online get-togethers, but many feel that people can only really communicate closely when meeting in person.

What is the potential and what are the limitations of online dialog? If we could see the boundary lines, could we then incorporate all the advantages of an online format into the process of public dialog in urban projects?

The difficulties of holding in-person public dialog were apparent in pre-pandemic times (at least to me and colleagues): We would struggle to gather together enough people; end up with the same kind of people turning up each time; and the results of the discussions, instead of being disseminated, would get buried. This will probably sound familiar to anyone who has been involved in public dialog. Online tools offer a new approach to public dialog, although they still require expertise and experience in in-person public dialog. What is the potential of this new approach?

11.3.2   Three Hypotheses on Online Public Dialog

Figure 11.7 shows three hypotheses regarding the potential of online tools in public dialog. The first is that online tools will make public spaces more open. With PCs, tablets, or other devices hooked up to the Internet, participants can attend meetings remotely. The second hypothesis is that online tools will increase the range of participation styles. By and large meetings can remain in person but provide alternative options for participation, such as allowing them to post text-based comments and ideas or allowing people to participate passively (by just observing the proceedings). The third hypothesis is that online tools can make the discussion process more transparent. When discussions are held digitally from the start, the time and effort of manually recording the proceedings are saved, the meetings can be uploaded to the web seamlessly, and the record of the discussions (such as who said what and when) can be published without any gaps.

Can these hypotheses be demonstrated in reality, and if so, under what conditions? To test the hypotheses, we organized a dialog with an online videoconferencing system with the support of Mimiguri, a creative consultancy with expertise in running workshops with online tools, in Ehime and Tokyo.

Online tools increase openness in public space	Online tools increase the range of participation styles	Online tools make the process more transparent

Offline

Meetings require people to get to the venue at the designated time, so only a few people can participate.

Offline

Most meetings have no variety in participation methods. Meetings are dominated by participants with the loudest voice, meaning that some participants never get a chance to contribute ideas.

Offline

Often, the results of the public dialog are not fully made public, making it hard to track the dialog's history.

Online

While people still have to attend at the designated time, they can attend remotely from their desired location, meaning that <u>more people will participate</u>. With no spatial constraints, hundreds of participants can potentially participate.

Online

Meetings can have more varied options for participation. Participants do not have to participate directly. They have the option of posting comments during the meeting or just viewing the proceedings, and they can also post comments at a later date. <u>The greater range of participation styles means that the "smaller voices" can also be heard.</u>

Online

Proceedings are recorded effortlessly and uploaded in real time, <u>enabling meetings to be continuously broadcasted far and wide. This makes the dialog process more transparent, increasing trust in</u> the project and encouraging greater participation and solidarity.

Fig. 11.7   Three hypotheses

11.3.3   Two Zoom-Based Workshops to Test the Hypotheses

The first online workshop we held was with people living around the University of Tokyo's Hongo Campus (Bunkyo, Tokyo). People were working or researching from home amid the pandemic, which would have had devastating ramifications for restaurants and shops around the campus. The topic for the meeting concerned the effects of the pandemic in the neighborhood. The online meeting was held using Zoom. The meeting was attended by four students from the University of Tokyo and 13 members of the public. The workshop could be accessed from nine prefectures, of which Nagasaki was the furthest away from the University of Tokyo (Fig. 11.8).

Each participant could play one of three roles: director, commentator, and audience. A director led the discussion. A commentator contributed ideas and engaged in dialog. An audience member listened as a third party and occasionally posted text comments on the chat box. This setup was designed to let participants choose a participation style that suited their personality and the degree of burden they were willing to shoulder. In a post-workshop questionnaire, all respondents agreed, and 90% agreed strongly, that having multiple options for participation style was valuable.

 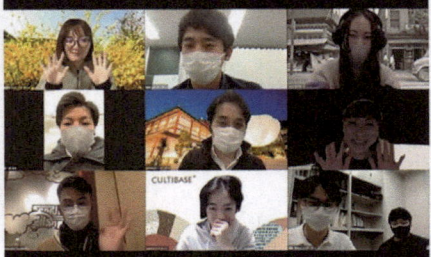

The participants view a street image provided by a member of staff on location.

Fig. 11.8 Scenes from the workshop on Hongo Campus

Itinerary for online workshop			
Intro	**Knowing**	**Creating**	**Summary**
Icebreaker, practice using tools	**Talk session about data-driven community building** Shin Osaki (a member of H-UTokyo Lab) held a dialog with researchers from Hitachi while referring to a CityScope demo session, which was organized by Mimiguri.	**Virtual walking tour with Google Street View** Each team had virtual walking tour to gain insights into Matsuyama City.	**Presentations**
Briefing on purpose and itinerary		**Research plan** With these insights, the teams pitched ideas for data collection in Matsuyama.	**Feedback from the researchers** The researchers commented on the presentations, suggesting ideas for what data should be collected and what precedents there are.

Fig. 11.9 Structure of online workshop with a stroll through Matsuyama

The second dialog was based in the central business district of Matsuyama City, Ehime. Dubbed an "online workshop with a virtual walking tour of Matsuyama," the workshop was designed so that participants could brainstorm ideas about Matsuyama's future while virtually strolling through the city on Google Street View (Fig. 11.9). This workshop was held on Zoom too, giving participants the choice to participate as they would in a regular workshop or just observe. Twelve participants participated actively and 27 observed. Around half of the participants were from Matsuyama City, but some participants were from outside Ehime Prefecture (some were from Kanto and Kyushu). One of the participants was an Ehime native who had attended university in Matsuyama City and who was attending a postgraduate course in Tokyo at the time of the workshop. The student said they wanted to find a job in which they will contribute to the future of Ehime.

The two workshops verified the hypothesis that online tools make the public space more open and the hypothesis that they increase the range of participation styles. The workshops were attended remotely by people interested in the topic, denoting a more open public space. The provision of different participation options meant that a greater range of people applied to participate, and the participants said themselves that having a range of options was valuable. Some of the participants said that they could listen to the meeting while doing housework, suggesting that online tools can enable people to participate even when they are busy with something else.

11.3.4   An Unsuccessful Attempt at Asynchronous Online Dialog

Both these examples of dialog used a synchronous format in which every participant had to access the same address at the designated time. The problem with this format is that the third hypothesis (that online tools increase the transparency of the discussion process) can only be established with an accumulation over time of publicized dialog outcomes. That is, the hypothesis can be established with asynchronous dialog. The work of testing the hypothesis can only proceed when there is a system for uploading the online discussions into the public domain to enable anyone to access them and enable the discussions to be extended or followed up.

To that end, we organized a series of three workshops and uploaded the discussions to our own public dialog platform. The workshops involved a quiz in which we presented vintage photos of Matsuyama streetscapes and the participants had to work out where the photos were taken. Perhaps because of this element of fun, spirits were high. Intrigued by the vintage photos, participants were inspired to discuss their thoughts about Matsuyama. We uploaded the discussions to the platform, expecting to spark asynchronous dialog. However, our expectations were confounded when we found that no one had left any comments on the platform.

Bewildered by the failure to achieve asynchronous dialog, we conferred with the workshop participants about the possible reasons. We identified two hurdles to participation in our platform site. The first hurdle was operational. The site had usability issues; it was unclear where to post comments and how to do so. The second hurdle was psychological: People felt shy about leaving a comment that would be exposed to the public eye. To lower the operational hurdle, we could make the site more navigable and provide a set of reaction buttons as an alternative to a text comment. The psychological hurdle could be addressed at different levels. At a superficial level, we could set questions that were easy to answer or offer an incentive (such as saying that the person's comments will be taken into account in the next workshop session). But to address the problem more substantially, we would have to give users a sense of who is in the conversation. In other words, the online platform would need to be linked to the scene of offline dialog so that participants can identify whom they are addressing rather than having to address an invisible and anonymous multitude.

11.3.5 Synchronous and Asynchronous Hybrid (Online–Offline) Dialog

Based on these assumptions, we have started (as of 2023) rebuilding our online platform with a Decidim[2] format and have attempted a dialog process in Ikegami (Ota, Tokyo) and Nomura (Seiyo, Ehime) involving back-and-forth between online and offline dialog (Fig. 11.10). At of time of writing, we are yet to test the platform directly, but we have gained the following findings in Ikegami.

In Ikegami, we ran a workshop and exhibition concerning local information, namely, vintage photos and information about local flora (Fig. 11.11). We also broadcast detailed information about the events on social media. Social media is a great tool for communicating information, but it also has similarities to an online platform in terms of the plethora of dialog and reactions and how these are publicized and logged. Thus, offline dialog (at the workshop and exhibition) was combined with asynchronous online dialog (social media), causing interaction between those engaged in dialog on each site. For example, we observed cases where someone who had participated in the exhibition later engaged in dialog on social media, and cases where someone engaged in social media (someone who has never met other participants in person and who has seldom attended a community-building event) later visited the exhibition and engaged passionately in dialog there. In these and other examples, we found that the range of people participating in dialog, and the range of dialog itself, enlarged with a back-and-forth between offline and online fora.

Fig. 11.10 Decidim homepage and webpage listing dialog topics (under construction) in Ikegami

---

[2]Decidim is an online tool used to collect opinion from a range of citizens and consolidate the discussion points so that they can be incorporated into the policymaking process. Meaning "we decide" in Catalan, Decidim was first developed in Barcelona in 2016 and was later localized into Japanese by Code for Japan. It is currently used by 400 organizations across 30 countries, with more than 900,000 participating. In Japan, it is used by Kakogawa, Shibuya, and other municipalities.

Fig. 11.11   Book exhibition in Ikegami

11.3.6   Hybrid Public Dialog, Not Just Online Public Dialog

In our initiatives so far, we have confirmed that using online tools for public dialog makes for a more open space and more diverse participation. People need not attend the meetings in person; they can attend remotely from the comfort of their home. They can also choose a level of participation, from actively participating to passively observing, according to their convenience and interest. Thus, online public dialog removes spatial constraints and gives participants choice; as such it can encourage participation from people who would not have participated in conventional in-person dialog. This means that public dialog has huge potential.

On the flipside, we also identified operational and psychological hurdles in asynchronous online dialog. When the public dialog starts and ends online, the psychological hurdle becomes too high for a successful dialog. It is therefore necessary to provide opportunities for offline dialog to make it more transparent as to whom people are dialoguing with. Essentially, there needs to be hybrid public dialog, integrating online and offline dialog.

11.3.7   Dialog with Citizens Unable to Meet in Person

In view of what we have gleaned about online public dialog, it seems that such dialog serves best in a supplementary role to offline public dialog by giving a voice to people who are unlikely to participate in person. These may be the voices of students who may have little chance to attend a meeting in person because of school or other classes, young people who moved away to attend university or a job but who care about their hometown and want to contribute to it, or working-age adults who are busy with a job or housework. Though it requires action to be taken to lower the psychological hurdle, asynchronous dialog offers great value in that it gives a voice to people with immensely important things to say about the future of the community.

Face-to-face discussions will always be important. But online tools can create opportunities to include in the dialog people who were unable to participate or who never really thought of doing so. Hybrid public dialog is an effective tool for increasing the range of people who can participate in the conversation.

11.4   Public Dialog in Data-Driven Urban Planning

In this chapter, I introduced attempts at data visualization and public dialog in the context of data-driven urban planning. Regretably, the social conditions did not allow these initiatives to be continuously put in practice as part of data-driven urban planning, but I shall end this chapter by discussing what can be anticipated when they are integrated into this cycle.

If a single package combining the data-visualization platform and public dialog platform existed, it could operate effectively alongside the urban data platform. While the urban data platform would focus on archiving and data linkage, this new combined platform would focus on facilitating communication between the stakeholders; it would serve as a digital communication platform.

In the case of Hanazono-machi Street, the data visualization and the dialog around it inspired ideas and feedback from local stakeholders, including calls for new data (such as data on takings in local shops) and an idea to install a permanent laser sensor. Essentially, the stakeholders wanted a new form of sensing and to make the community more data-driven. When data are presented right, they hold up a mirror to the community. Stakeholders will see what is going on in their community and then spontaneously start coming up with ideas for making improvements. This suggests that a data-driven city is not just led by smart city operators but also by the community. We have seen the rise of a kind of community-level, grassroots leadership, which we should perhaps dub data-driven local management. Data visualization done right will spark such grassroots community leadership.

We also saw clear evidence that a hybrid approach to public dialog can throw open the doors to people who were unable to participate in conventional in-person workshops. Our workshops were attended by graduate students, housewives, and other people who were participating in community building for the first time, via social media. Hybrid dialog increases the range of participation and thus the range of perspectives represented.

From our observations about data visualization and public dialog, we concluded that combining the two into a single platform (a digital communication platform)

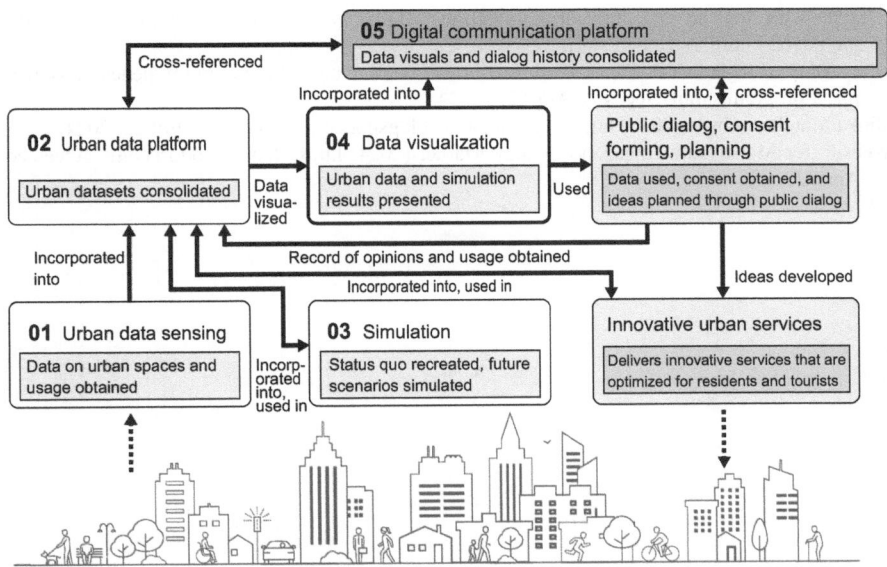

Fig. 11.12 Updating data-driven urban planning

will help diversify the range of people participating in data-driven local management (community-led management), bringing in a wider array of perspectives and opinions. Under these conditions, public dialog will go beyond the task of simply gaining public consent toward a proposed intervention and become a step toward a more creative and more grassroots form of community building (Fig. 11.12).

References

Barcelona City Council (2024) Superilles. https://ajuntament.barcelona.cat/superilles/en. Accessed April 22, 2024.

Maipure (Matsuyama, Iyo, To'on, Matsumae, Tobe) (2023). https://matsuyama.mypl.net. Accessed April 22, 2024.

Matsuyama City (2018) *Aruite kuraseru machi Matsuyama no arata na shinboru rōdo Hanazono-machi dōri rinyūaru nigiwai to kōryū o hagukumu hiroba o sonaeta dōro* [A new symbol of a walkable, livable Matsuayama: Revamping Hanzaono-machi Street, creating a buzz and community engagement]. March 2018. https://www.city.matsuyama.ehime.jp/shisei/kakukaichi-ran/tosiseibibu/dourokensetuka.files/300305hanazono_A3panfu.pdf. Accessed April 22, 2024.

Ministry of Land, Infrastructure, Transport and Tourism (2019) *Matsuyama sumātoshiti purojekuto (Matsuyama sumātoshiti consōshiamu)* [Matsuyama Smart City project (Matsuyama Smart City consortium)]. https://www.mlit.go.jp/toshi/tosiko/content/001514520.pdf. Accessed April 22, 2024.

Ministry of Land, Infrastructure, Transport, and Tourism (2020) *Shingata koronawirusu kansenshō ni taiōsuru tame no endō inshoku ten tō no rojō riyō ni tomoau dōro senyō ni tsuite* [On occupation of roadside areas connected with use of street by roadside restaurants and the like to deal with the novel coronavirus infections]. https://www.mlit.go.jp/report/press/road01_hh_001624.html. Accessed April 22, 2024.

NYC DOT (2024) NYC Plaza Program. https://www.nyc.gov/html/dot/html/pedestrians/nyc-plaza-program.shtml. Accessed April 22, 2024.

Ojoka Marche website (2023) https://ojokamarche.blogspot.com/. Accessed April 22, 2024.

Website for Matsuyama Hanazono Sunday Market (2024) https://hanazonodori.com/. Accessed April 22, 2024.

Chapter 12
Project 2: Smart Aging

Katsuya Iijima, Ken Naono, Yoshinori Sato, and Tadashi Mima

Abstract This chapter discusses Japan's efforts to address the challenges encountered by its super-aging society through an AI-driven frailty-prevention program. Long life expectancy in Japan coupled with its declining working-age population has increased healthcare expenditures, thus emphasizing the necessity for preventive measures. The concept of "frailty" is introduced as an aging-related decline in physical and mental abilities (capacities), which can be reversed through appropriate interventions. A holistic and comprehensive approach to frailty prevention that addresses nutritional, physical, and social factors is proposed. The Institute of Gerontology at the University of Tokyo has developed a frailty screening program for community-dwelling older adult volunteers in Kashiwa City. The program uses simple tests to identify the risk of sarcopenia and encourages social participation. Data from these screenings, combined with other health datasets, are analyzed using AI to deliver personalized recommendations for preventing frailty. This initiative aims to integrate digital technology into community life, thus fostering a sustainable "smart-aging society". This chapter concludes by emphasizing the importance of evidence-based, resident-led programs and the necessity for robust data governance

K. Iijima (✉)
The Institute of Gerontology/The Institute for Future Initiatives, The University of Tokyo, Tokyo, Japan
e-mail: iijima@iog.u-tokyo.ac.jp

K. Naono
The Digital Platform Innovation Center, Research & Development Group, Hitachi, Ltd., Tokyo, Japan
e-mail: ken.naono.aw@hitachi.com

Y. Sato
The Design Center, Research & Development Group, Hitachi, Ltd., Tokyo, Japan
e-mail: yoshinori.sato.uw@hitachi.com

T. Mima
Smart Infrastructure Consulting Department, Hitachi Consulting, Tokyo, Japan

Graduate School of Media and Governance, Keio University, Kanagawa, Japan
e-mail: tmima@hitachiconsulting.co.jp

© The Author(s) 2025
Hitachi-UTokyo Laboratory (H-UTokyo Lab.), *The Architecture of "Society 5.0"*,
https://doi.org/10.1007/978-981-96-2929-9_12

to realize a human-centric, multigenerational community within the framework of Society 5.0.

Keywords Frailty prevention · Super-aging society · Health datasets · Explainable AI (artificial intelligence) · Data governance

12.1 The Need for Frailty Prevention

This chapter introduces efforts to meet the challenges of Japan's super-aging society, which is becoming increasingly socially cohesive for all generations. Such a society empowers people to remain constructive members of society throughout their lives; it promotes independence and social participation, and provides comfortable living environments along with the assurance that one can continue to age in place even after growing frail. The main initiative introduced here is an AI-driven frailty prevention program run in Kashiwa City, Chiba prefecture. This section describes the challenges encountered in this project and their outlook.

With one of the highest life expectancies in the world, Japan has become a country of long lifespans, something for which humankind longs. This is a testament to the quality of Japan's healthcare system (in 1961, Japan introduced a national health insurance for all and advanced medical technologies) and an excellent public health policy. However, longer lifespans have created a super-aging society unprecedented in human history. As the Japanese population continues to age, population projections (across 5-year time points from 2020) suggest that the older adult population (aged 65 or older) will increase more slowly than the working-age population. The projections also suggest that an increase in people dependent on nursing care will cause nursing care expenses to surpass the 2018 figure of 10 trillion yen and reach 24.6 trillion to 25.8 trillion by 2040 (Cabinet Secretariat of Japan et al. 2018). In light of these projections, the Ministry of Health, Labour, and Welfare released a white paper in 2022 with recommendations about "the era of 100-year lifetimes." The white paper highlighted the need to improve individuals' quality of life, employment, and social participation throughout life and to suppress the rise in the number of care-dependent people (Ministry of Health, Labour and Welfare 2022).

Following the publication of these recommendations, in 2014 the Japan Geriatrics Society proposed adopting the English word "frailty" (as the loanword *fureiru*) to introduce a new perspective into policymaking for promoting wellness and preventing long-term care dependency and to make the public more mindful of the need to look after their own health (The Japan Geriatrics Society 2014). As a clinical condition, "frailty" means an aging-related decline in physical and mental abilities that diminishes the person's ability to withstand, or recover from, stressors. As illustrated in Fig. 12.1, frailty is a transitional stage between robust health, disability, and dependency. When a person is frail, a variety of physical, mental, and social factors can combine in a vicious cycle, increasing their vulnerability to adverse health outcomes such as debilitating conditions or death. However, frailty occurs

Frailty is an aging-related decline in physical and mental abilities

1. Frailty is a transitional stage between robust health and dependency.
2. Frailty involves a combination of factors (with psychological and social factors affecting physical decline).
3. Frailty is reversible; it is a period in which the person's functions can be restored with the right intervention.

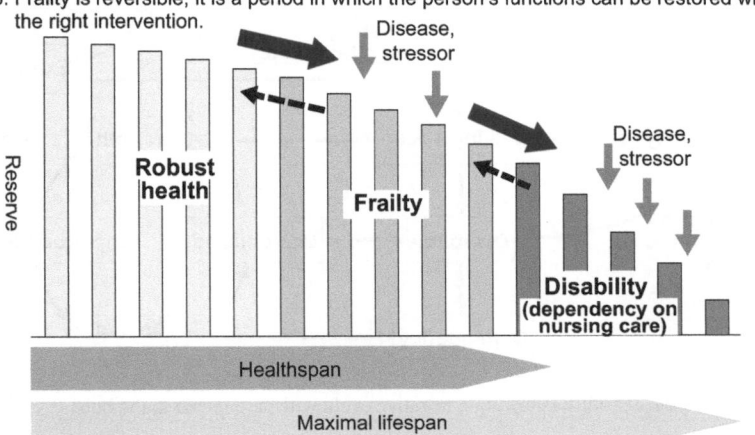

Fig. 12.1 The concept of frailty. (Figure adapted from Kuzuya 2009)

before the irreversible debilitation stage; therefore, the condition can be reversed by appropriate intervention. Thus, the frailty concept implies that there is a need to create new evidence-based interventions in communities to improve health.

Therefore, a holistic and comprehensive approach is required to effectively prevent or treat frailty. This approach must be explained by understanding the key role of chronic undernutrition in frailty (sarcopenia is a major cause of frailty). Figure 12.2 is adapted from Fried and Walston's cycle of frailty (Fried et al. 2001), which represents the factors of frailty as part of a negative cycle. Notably, rather than just focusing on managing a particular disease or injury as in a conventional medical model, the cycle of frailty focuses on the key symptoms of frailty, including sarcopenia-associated physical decline (reduced walking speed, immobilization, and increased risk of falls and injuries), unaccountable fatigue, and diminished activity. Thus, the core components of the frailty model are factors that facilitate activities and wellness in everyday life (participation in community affairs and social interactions).

The vicious cycle described in the cycle of frailty is not only dependent on physical decline; it is also shaped by psychological and social factors (lack of community participation, social networks, and social support), according to recent studies I conducted with colleagues at the University of Tokyo (Tanaka et al. 2022; Lyu et al. 2022). Studies have shown that the best way to prevent frailty and dependency is to adopt a holistic and comprehensive approach to health assessments and individually tailored interventions.

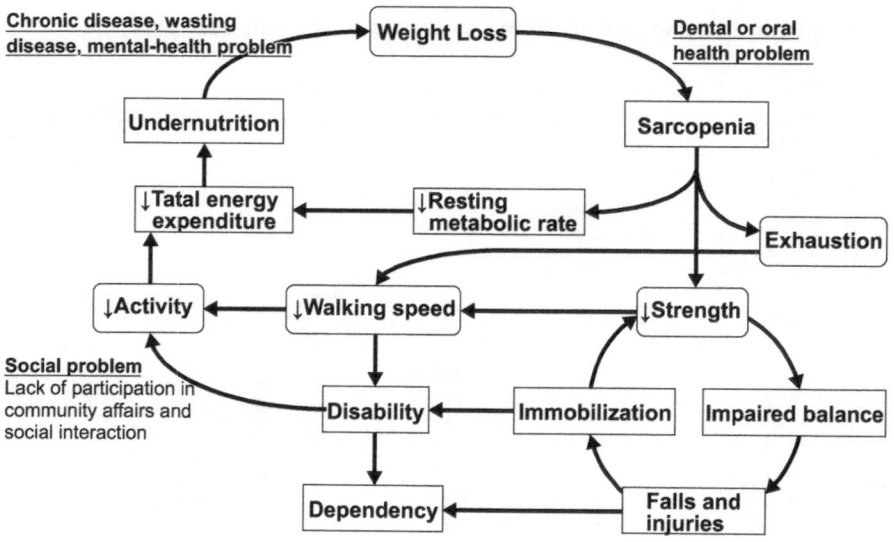

Fig. 12.2 The cycle of frailty (showing a negative cycle with sarcopenia at the hub)

Thus, the time has come to take the first step in supporting longer health spans; businesses, academics, and the government should come together and devise, using the latest available evidence, strategies for addressing the three key factors of frailty risk: nutritional (diet and oral health), physical (everyday activities and exercise), and social (community participation). With the pandemic encouraging people to stay indoors and withdraw from community activities, we are witnessing pandemic-associated frailty: people who, as a result of being inactive, develop severe sarcopenia, leading to secondary health damage. We have to go beyond conventional healthcare policies by further integrating digital technology into community life, so that people of all generations have more opportunities to socialize and participate in the community. We also need a genuine population approach that uses a blend of self-help and mutual support to create communities where people are empowered to undertake wellness support activities.

12.2   The Need to Supplement Conventional Health and Wellness Interventions with Programs for Encouraging Social Participation in Older Adults

Figure 12.3 shows the statistics from a Cabinet Office white paper on aging (Cabinet Office 2022). Statistics suggest that lifespan will increase to over 90 years for women and 84 years for men by 2050. It is nice to know that people will live longer, but it is a concern that the population will continue to shrink and age (the population gets smaller overall and has a greater share of older adults [aged 65 or older]) as we

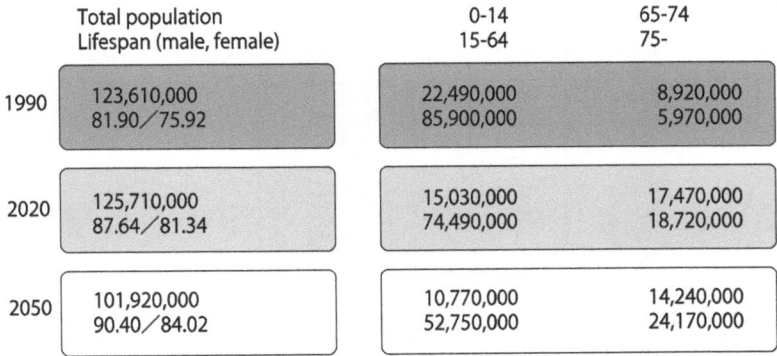

	Total population Lifespan (male, female)	0-14 15-64	65-74 75-
1990	123,610,000 81.90／75.92	22,490,000 85,900,000	8,920,000 5,970,000
2020	125,710,000 87.64／81.34	15,030,000 74,490,000	17,470,000 18,720,000
2050	101,920,000 90.40／84.02	10,770,000 52,750,000	14,240,000 24,170,000

Fig. 12.3 Japan's past population trends and population projections

move toward 2055. Of particular concern is the growing population of late-stage older adultse (aged 75 or older).

With over 75s making up a growing share of the population, it has become even more urgent to provide early health and wellness interventions to prevent frailty. Over the years, the government's wellness interventions have traditionally focused on raising public awareness about having a balanced diet with plenty of protein and regular exercise. However, new evidence suggests that this approach needs to be updated. Residents should be given chances to reassess their everyday activities from a holistic perspective and with a view to adopting activities that will increase their healthspans; in light of the latest evidence, people should be encouraged to engage in social activities in the community, rather than just be encouraged to eat a healthy diet and exercise regularly. Local governments can promote health-building that allows elderly people to step forward toward social participation and activities to contribute to communities. In short, a community-building approach, with residents taking the lead in building and protecting their own communities, will become an increasingly important part of promoting wellness.

12.3   Pioneering Frailty Prevention in Japan

How can we prevent frailty, encourage older adults to engage in the community, and create opportunities for them to lead productive lives? The Institute of Gerontology in The University of Tokyo developed a frailty screening program based on evidence obtained from a lengthy period of longitudinal observational cohort study for community-dwelling elderly people (so-called the Kashiwa Study). In the screening program, community-dwelling elderly people are trained to act as volunteers (dubbed "Frailty prevention supporters" in Japanese-English) who lead efforts to screen fellow residents for frailty risk. These resident-led screenings use a holistic approach that considers nutritional (diet and oral health), physical (exercise and non-exercise life activity), and social factors (social participation and social

Finger-ring test

When you try to form a finger ring around your calf, which of the following applies?

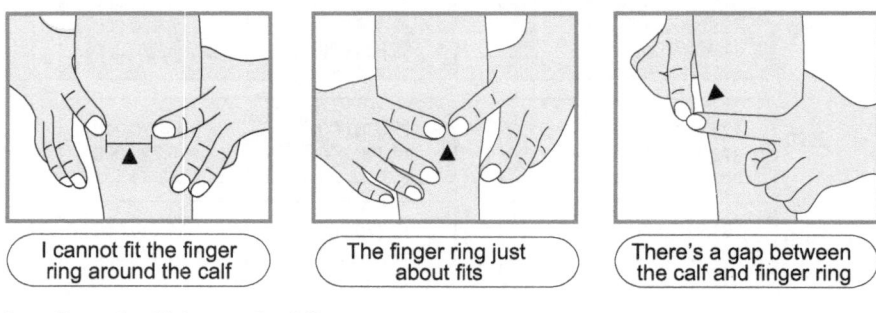

I cannot fit the finger ring around the calf	The finger ring just about fits	There's a gap between the calf and finger ring

Questionnaire ("eleven-check")

Please answer the following questions by affixing a sticker of the appropriate color onto the appropriate answer box. Note that the position of "yes" and "no" answers is reversed in the shaded question items. Make sure you affix the sticker on the correct box.

Nutrition	1. In comparison to your peers of the same sex, do you take care about your diet?	(Yes) (No)
	2. Do you eat both vegetables and protein sources (meat from warm-blooded animals, meat from fish and other seafood) at least twice a day?	(Yes) (No)
Oral health	3. Are you usually able to chew hard foods like dried shredded squid or pickled radish?	(Yes) (No)
	4. Do you sometimes misswallow liquids like tea and soup?	(No) (Yes)
Exercise	5. Have you engaged at least twice a week, over the course of at least a year, in exercise sessions that last at least 30 seconds and in which you break a sweat?	(Yes) (No)
	6. As part of your everyday activities, do you walk or engage in similar exercise for at least an hour a day?	(Yes) (No)
	7. Do you walk faster than your peers of the same sex do?	(Yes) (No)
Socio-psychological	8. Do you go out less often than you did last year?	(No) (Yes)
	9. Do you share a meal with someone at least once a day?	(Yes) (No)
	10. Do you feel energetic?	(Yes) (No)
	11. Are you worried above all that you are forgetful?	(No) (Yes)

Fig. 12.4 Frailty screening test (simple self check)

network) associated with frailty. The purpose of the program was to encourage elderly people to reflect on their everyday activities in view of the multiple dimensions of frailty (multi-faceted frailty) and to adopt attitudes and behaviors that will help prevent frailty.

Figure 12.4 shows the screening test. Among frailty screening tests, this is relatively simple, making it practical for self-screening.

The test has two parts. The first was a self-measured "Finger-ring test (Yubi-Wakka test; in Japanese)" used to assess skeletal muscle volume related to progression of frail state simply. The person joins their two thumbs and two index fingers to form a ring. The size of the ring remains largely constant throughout a person's life, with an inner circumference of approximately 30 cm, which is proportional to the person's height. The person forms this finger ring around the thickest section of their nondominant calf (without overexertion) to compare the circumference of this section of the calf with the circumference of the finger ring. If the calf is too large for the finger ring or if the finger ring can only be formed around the calf, the person affixes a blue sticker onto the sheet. The blue sticker indicates a favorable result; the person is not at risk of sarcopenia. If there is a gap between the calf and finger ring, the person affixes a red sticker onto the sheet, indicating that they are at risk of sarcopenia and should be cautious. We adopted this Finger-ring test as the program's main screening check because it is simple. By simply measuring muscles in the calf, sarcopenia risk in the whole body can be screened.

The second part of the test consists of a self-administered questionnaire dubbed the "eleven-check." The questionnaire had 11 questions covering four domains of everyday activities (nutritional, oral, physical, and social/mental/cognitive). Examples of the varied questions include "Do you eat both vegetables and protein sources (meat from warm-blooded animals, meat from fish and other seafood) at least twice a day?" "Do you walk faster than your peers of the same sex do?" "Do you share a meal with someone at least once a day?" For each question, the person affixes a blue or red sticker to indicate an affirmative or negative response, respectively. In this way, the person identifies aspects of their everyday life to improve and starts taking a greater personal interest in preventing frailty.

These two tests constitute a brief version of the screening. The full screening program included 22 components and takes an hour or two to complete. The items included physical tests, such as the chair stand test by one-leg, grip test, measuring skeletal muscle volume and oral diadochokinesis (quantifying smoothness of speech) test.

Figure 12.5 shows photos of the tests conducted during the full screening program. The residents acting as volunteers create a yellow-green uniform (standard

Residents across Japan conducting the full screening program
Photos used here with permission from:

Fig. 12.5 Full screening test in action (volunteers working to screen fellow residents for frailty). (Photos used here with permission form: Institute of Gerontology, The University of Tokyo and Iijma K, *Fureiru yobō handobukku* [Frailty-prevention handbook])

across Japan) and administer the program themselves, giving them a life-affirming feeling that they are contributing to the community. In this way, the residents discover issues in their own lives that they need to address, mutually enhance each other, feel empowered to take self-supporting actions to address these issues, and feel empowered to help other residents reflect on their health. This concept of convivial mutual support is central to the program, and, as the smiles in the photos reveal, the residents enjoy the process. The program represents an attempt to use screening tests based on the latest evidence to empower residents to monitor their own health and each other's health through their own efforts. In addition, thei activity is also a new system of self-help and mutual aid for a new era.

The resident-led frailty prevention program began in Kashiwa City. As of 2022, it has been rolled out in 96 municipalities in Japan. All screenings were performed with the consent of each participant. The screening data were stored by each municipality, and at certain intervals, were uploaded to a control center at the University of Tokyo with anonymized personal data.

12.4   A Task to Be Addressed to Enable the Use of Data for Frailty Prevention

As mentioned above, data obtained from the full screening program were logged, providing an important source of basic data to guide efforts to prevent frailty. However, a major problem exists with the use of data relevant to frailty prevention. Local governments maintain many different databases relevant to wellness promotion, including residents' health check data, medical claims data, drug prescriptions, and data on elderly nursing care services used. These databases, which cover all residents, have remarkable potential value. The problem, however, is that the value realized by using databases held by the local government has been limited in practice; they are never fully used, and residents are unable to see how they can prevent frailty, what actions should be taken now, and where they should start.

In view of this problem, we identified a task that should be addressed when it comes to a frailty prevention strategy, which needs to be integrated and analyzed with AI-driven analytics to identify the relevant information to present to each resident (issues they need to prioritize or actions they are recommended). In other words, the task was to deliver AI-driven frailty prevention, with AI analyzing the datasets in an integrated manner to deliver a bespoke set of recommendations to each resident. For one resident, the AI might identify recommend oral care; for another, it might recommend both oral care and greater participation in community affairs; and for another, it might recommend a more balanced diet. Guidance and support will vary depending on the person's circumstances and lifestyle.

Figure 12.6 compares the AI-driven frailty prevention with conventional strategies for preventing care dependency. True, the conventional approach is data-driven to a certain extent and, in that sense, it is evidence-based. However, rather than

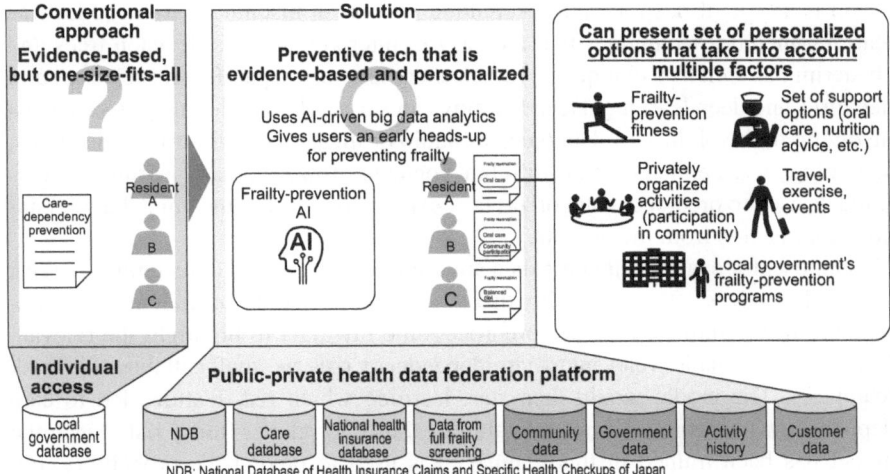

Fig. 12.6 A new frailty-prevention strategy that uses public-private health data federation platform health data

delivering personalized guidance, it delivers blanket guidance to all users, such as pointing them toward exercise classes designed to help prevent care dependence (see the left part of the figure). Such one-size-fits-all guidance will help prevent care dependency, but its effect is limited because it fails to account for the diversity in elderly people's lifestyles and thus fails to provide personalized support.

By contrast, the proposed AI-driven approach is based on evidence and delivers personalized guidance (see the middle of the figure). I shall describe the approach in more detail in the next section; however, the basic concept involves big data analytics, whereby the AI program can match data with the individual's needs to give the person an early heads-up on frailty risk. This implies that every user receives a set of recommendations tailored to their particular situation and needs (see the right side of the figure). Each local community will have a public-private framework linking a mixture of frailty prevention services: public, private, payable, and nonpayable. When an AI program identifies frailty risks in a user's everyday life, the user can be referred to local community services that address these risks, enabling effective frailty prevention.

12.5   Frailty-Prevention AI

The frailty prevention AI is described in this section. It comprises of two technological components. The first component is an explainable AI called the Black Box Breaker (B3), developed by Hitachi. The second component is technology for managing evidential data.

B3 is a type of deep learning technique popular in machine learning. Machine learning includes techniques that are easy to interpret, such as decision trees and clustering, as well as techniques with excellent learning capabilities, such as neural networks and deep learning. Deep learning can surpass human learning in problems involving high volumes of data (such as audio or image recognition). In the 2010s, technology was applied to many different contexts. However, many learning models (models used to predict or categorize) lack explanatory power, meaning that we cannot gain insights from the outputs.

Suppose, for example, that big data are used to train a model to recognize the risk of care dependency (the probability that a person will become dependent upon nursing care in the future). First we preprocess the raw data to highlight the relevant features (the variables relevant to care-dependency risk; we shall call these the "risk features"). The model would then start learning which risk features led to care dependency and which did not, to gain the ability to predict frailty risk. Once the model has been trained in this manner, it is used to predict the probability. For example, the model may predict that Person A 56% chance of becoming care-dependent within 3 years. If the training datasets are sufficient, the model would be highly precise in its predictions. However, while the model has predictive power, it may lack explanatory power; it may not be able to tell exactly which risk features are based on its predictions. If it lacks explanatory power, it will not be possible to identify insights from predictions for developing personalized guidance to prevent the person in question from becoming care-dependent.

However, some deep learning models have explanatory power; they include an explanatory mechanism so that we can determine which risk features shape the outputs. B3 is among them. It is adept at learning from small samples and high-level datasets, can present its interpretations in a statistical model, and when making predictions, can provide an account of the features from which the predictions are derived. Applying these abilities to frailty prevention, B3 could give a percentage probability that the elderly person in question will become care-dependent and also show the risk features underlying such a probability, so that we can see the evidential basis for the prediction.

In addition to using B3 to give a probability percentage of care dependency along with the predictor variables underlying this prediction, frailty prevention AI can trace the risk features themselves back to the evidential data underlying them. Figure 12.7 shows this regression. The model tells us that Person A 56% probability of becoming care-dependent. This also indicates that the predominant risk feature underlying this prediction is oral frailty risk. It then goes back further by citing the evidential data for this risk feature: Person A reported during a frailty screening at a given time point that they had difficulty chewing pickled radishes.

Thus, our frailty prevention AI uses evidential data management to track the relationship between datasets, risk features, and the risk prediction model. Therefore, the outputs of the predictive model can be traced to the raw data.

Figure 12.7 illustrates this. This model has been identified as a risk factor of oral frailty. It can also cite the evidence for this: During a past frailty screening, the person answered "No" to the question, "Are you usually able to chew hard foods like

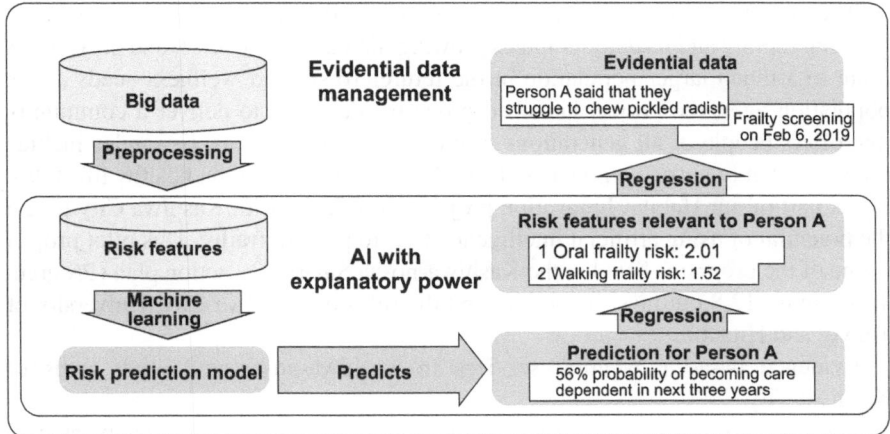

Fig. 12.7 Frailty-prevention AI with example of how it can be used

dried shredded squid or pickled radish?" or the person's medical invoice stated, as name of the condition, "diminished oral function."

The system quantitatively represents the relative weight of each risk feature when predicting and ranking them. Because a high-ranking risk feature can be traced to the raw data, we can, in the case of Person A, see that the prediction was grounded in the fact that Person A reported that they had difficulty chewing pickled radish in a frailty screening.

Thus, the frailty prevention system combines a model with the explanatory power of B3 with evidential data management and, as such, can collate data in the national health insurance database with frailty screening responses to derive frailty prevention strategies tailored to a person's lifestyle, situation, and needs. Thus, it can help nursing care experts make better decisions, and serve as a reference resource for expert panels.

B3 has already been used as a biomarker discovery service for pharmaceuticals that automatically searches for key predictors of drug efficacy. In 2020, the service earned a Good Design award in Japan in recognition of how it facilitated communication between AI and its users (Hitachi 2019).

12.6   Feasibility Study of Frailty-Prevention AI in Kashiwa-no-ha Smart-City Project

Many Smart-City Projects are developing data-driven solutions to address city challenges. One such project is the Kashiwa-no-ha Smart City in Japan. The project was launched in 2011, and a 3-year action plan was formulated in March 2020 (Kashiwa-no-ha Smart City Consortium 2021), with four areas set for data-driven solutions: mobility, energy, public spaces, and wellness.

The area where the Kashiwa-no-ha project is underway has seen its population growing rapidly and its demographics growing increasingly diverse, and projections point to a much larger population in the future. To support wellness needs as the population ages, the project launched programs designed to deliver a community that offers people of all generations future health and wellness. Examples include programs that focus on helping people of all generations maintain healthy lifestyles.

As part of our Habitat Innovation project, we worked with Kashiwa City to test the potential of using artificial intelligence (AI) to prevent frailty. This pilot project is one of the programs listed in the Kashiwa-no-ha Smart City action plan (Program 4.2). Figure 12.8 outlines the project and the roles of Kashiwa City, University of Tokyo, and Hitachi.

Evidence based prevention services in Kashiwa-no-ha and other areas of Kashiwa City.

The program involves the following steps: AI (B3) collates and analyzes data from datasets relevant to frailty prevention, including health checks, medical claims, drug prescriptions, and data on elderly nursing care services. Stored on the database called "Kokuho database (KDB)," operated by public insurers in each prefecture, and the data from resident-led frailty screening (developed by Iijima K, IOG in the University of Tokyo). AI also maps data to a person's health status, lifestyle interests, and preferences. Having done so, AI presents a set of bespoke recommendations for improving wellness and preventing care dependency. Consequently, the services that a person receives will be more effective than the current services intended to prevent frailty. We launched this program because we felt that conventional approaches failed to identify high-risk cases or deliver effective interventions owing to a lack of integration between the medical side (medical checkups and medical treatment related to frailty prevention) and the everyday activities side

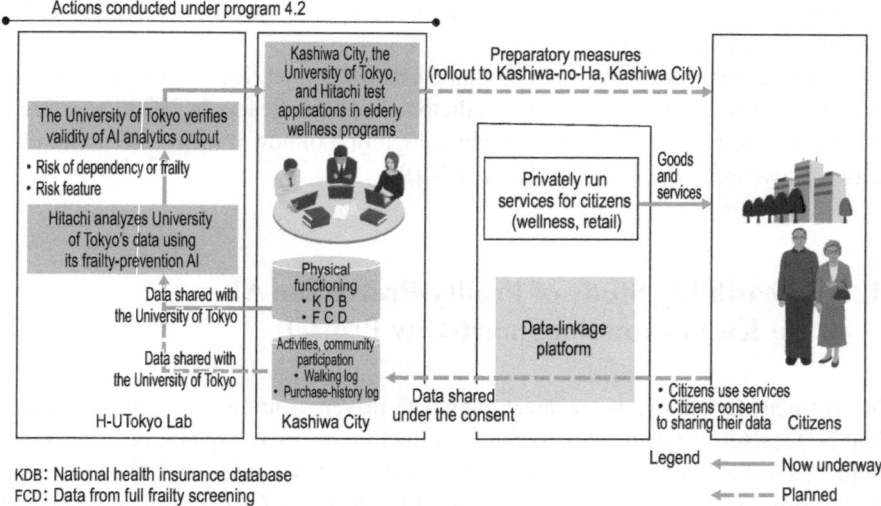

KDB: National health insurance database
FCD: Data from full frailty screening

Fig. 12.8 Outline of project for testing the potential of using AI to prevent frailty

(encouraging the person to sustain activities that would help prevent frailty). We also felt the need for an effective population approach that would encompass a greater number of citizens. These elements need to be further aligned to encourage target residents to modify their behavior in a sustained manner.

At the time of writing, the process of vetting and anonymizing Kashiwa City's personal data was complete, and as part of the Habitat Innovation project, we started AI analytics.

12.7 Datasets Used in the Kashiwa-no-ha Smart-City Project

The dataset described in the previous section is used to train the AI model. The dataset amounted to big data, covering some 110,000 elderly residents (aged 65 years or older) of Kashiwa City. The number of features converted from the dataset was approximately 900.

Our system analyzes these data using the B3 described above (see Sect. 12.5) to identify factors related to frailty. Factors correlated with frailty include care-dependency designation, health inspection results, and underlying diseases.

Before the AI is trained on big data, the raw data must be properly prepared. Part of the pre-processing involved 900 features. In the case of medical inspection results, for example, raw figures will be of little use to the model. They need to be preprocessed, usually by setting absolute thresholds, so that the model understands which values indicate that caution is required. Figure 12.9 illustrates such pre-processing with an example of body mass index (weight in kilograms divided by the square of height in meters). The table in the figure uses a more granular set of ranges than those presented by WHO.

Feature	Absolute threshold
Underweight 1	BMI = < 16.0
Underweight 2	BMI = 16.0–16.9
Underweight 3	BMI = 17.0–18.4
Normal weight 1	BMI = 18.5–19.9
Normal weight 2	BMI = 20.0–22.9
Normal weight 3	BMI = 23.0–24.9
Obesity class I 1	BMI = 25.0–27.4
Obesity class I 2	BMI = 27.5–29.9
Obesity class II	BMI = 30.0–34.9
Obesity class III	BMI = 35.0–39.9
Obesity class IV	BMI = ≥ 40.0

Fig. 12.9 BMI grade

We delineated three underweight categories (whereas the WHO uses only one category for underweight) because having multiple underweight ranges and tracking variations over time within these ranges can help in preventing frailty. When labeling diseases, we used ICD-10 codes (Ministry of Health, Labour and Welfare 2021). We did so because this is the standard disease classification, because it will help medical professionals evaluate the efficacy of the AI outputs, and because it means that any rare disease that could be used to identify an individual can be consolidated into a group of diseases or erased.

The machine-learning model learns the objective variables (variables representing a particular phenomenon) and explanatory variables (variables explaining that phenomenon) as a set. An example of an objective variable is whether a person at a given point in time has a dependency grade of one or higher. There are seven grades of dependency. The first two grades indicate that the person requires assistance, and the other five grades indicate dependency on nursing care, with a higher grade indicating a greater level of dependency. The explanatory variables paired with this objective variable consisted of data from the frailty screening. In this case, the machine learning model understands that a person is at risk if they have more than a certain number of red stickers on their frailty screening sheet and can also pinpoint the particular components of frailty screening that suggest risk. This information is then relayed to the person in question. The person will therefore have a clearer idea of their health status and what specific actions they should take to prevent frailty. Eventually, it will be possible to determine whether the person ultimately averts the risk of becoming care dependent.

One of the major issues is verifying the validity of AI outputs and guidance. Rather than taking the outputs and guidance at face value, we need to check whether they intuitively make sense and broadly align with the types of decisions typically made by healthcare professionals. To this end, the University of Tokyo's Institute of Gerontology will organize panels comprising gerontologists and experts involved in healthcare and nursing policymaking.

12.8 Obtaining Consent Toward the Use of the Data

Chapter 6 ("Data Governance") discusses how parties receiving data need to comply with legal and regulatory requirements and build governance that exceeds these requirements, and how this requires the receiving parties to obtain consent from stakeholders and design a process for handling the data.

When obtaining data from frailty screening, the first step is to ensure that residents consent to use their data for research purposes. Under the latest action plan for the Kashiwa-no-ha Smart City, Kashiwa City will link screening data with relevant data from the National Health Insurance Database and share the linked data with the University of Tokyo's Institute of Gerontology. Before it does so, Kashiwa City will apply to the Kashiwa City council for the protection of personal data (a body established pursuant to a Kashiwa City ordinance on personal data protection) and to the data publication and personal data protection committee of the Chiba Prefecture Wide Area Union for the Medical Care System for the Elderly (a governmental organization that is responsible for public health among elderly residents of Chiba Prefecture, which entrusts Kashiwa City with operations involving the national health insurance database); to obtain permission to share anonymized data as a practical and effective means for serving the program for integrated public health and care-dependency prevention for elderly people, a program was launched in the year that ended in March 2021.

As part of the application process, we consulted Kashiwa stakeholders to obtain their consent for the use of the data. As mentioned in the previous section, this consultation process will focus on the new value and data applications that can be generated from frailty-prevention AI, but it will also involve expert panels to consider the findings of AI analytics and the elderly care services that could be implemented in Kashiwa so that stakeholders can understand the specific processes for the implementation and delivery of services to residents.

Next, as mentioned in the previous section, Kashiwa City shared this information with the University of Tokyo after anonymization. To anonymize the data, Kashiwa hashes them (by mapping them to a randomly selected and non-invertible value) and removes identifiable information (such as names and addresses), leaving only data items necessary for the analysis. This arrangement ensured that only Kashiwa City could link the results of the analysis performed on the anonymized data with the individuals. The University of Tokyo analyzed the data within the scope of the purpose specified in the action plan for Kashiwa-no-ha Smart City. In practice, Hitachi runs AI-driven analytics on behalf of the University of Tokyo and within the scope the university specifies. To ensure data security, Hitachi re-anonymizes the data it analyzes by converting discrete age data into 5-year age groups and removing individuals with rare diseases (diseases that occur in fewer than ten cases) from the data. This re-anonymization is designed to minimize the risk of an individual being identified from data or analytic outputs. As such, it is an example of governance that goes beyond the minimum safety criteria specified in the law and relevant standards.

The system was designed to prevent data from leaving the data center of the University of Tokyo. Kashiwa City and the Chiba Prefecture Wide Area Union for the Medical Care System for the Elderly checked the security of the data center. Another feature of the system is that the University of Tokyo screens data-driven research from an academic perspective to ensure that the research outcomes benefit the community in Kashiwa. Once the research was completed, the data were destroyed in accordance with an agreement signed by Kashiwa City and the University of Tokyo.

Fig. 12.10 shows the roles of the parties who handle the data.

The parties are arranged horizontally, whereas their roles in the data are arranged vertically. "Supporters" refers to the resident volunteers who lead the frailty screening. "Union" refers to the Chiba Prefecture Wide Area Union for the Medical Care System for the Elderly. "Chiba health insurance" refers to the Chiba National Health Organizations. The "KDB" refers to the Kokuho (National Health Insurance) database and is operated and managed by the Chiba National Health Organizations. The

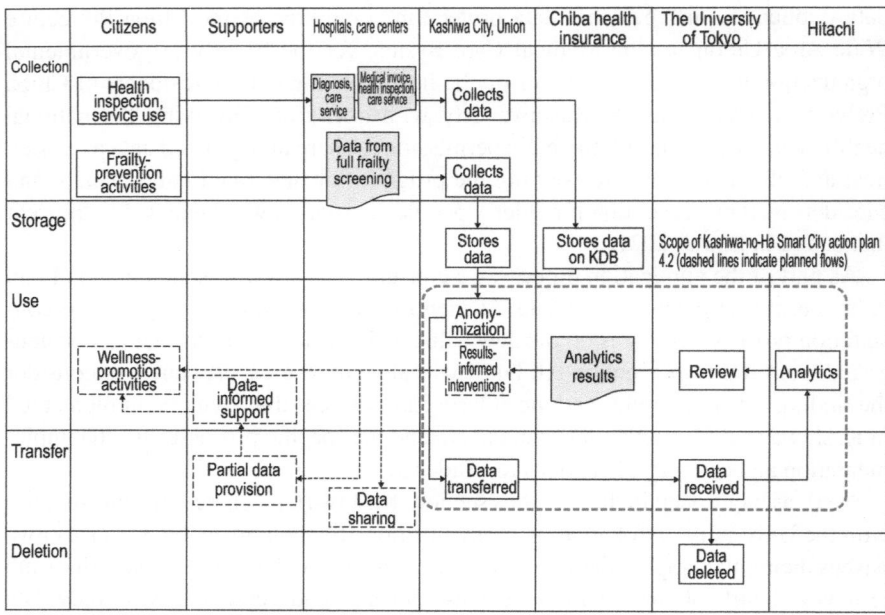

Fig. 12.10 Data flows and roles of parties

figure indicates that the data were transferred to the University of Tokyo as part of Program 4.2. The university received these data from Kashiwa City. Kashiwa City obtains data from the National Health Insurance database of the Chiba National Health Organization pursuant to an agreement between the Kashiwa City Council for the Protection of Personal Data and the Chiba Prefecture Wide Area Union for the Medical Care System for the Elderly.

"Results-informed interventions" and "wellness-promotion activities" refer to personalized sets of guidance that take into account the array of factors pertaining to the person in question (see Fig. 12.6). "Data-informed support," "partial data provision," and "data sharing" refer to the use of the analytics results by and between Kashiwa City and the screening volunteers or by and between Kashiwa City and health service providers. The data flows indicated by dashed arrows remain in the conceptual stage at the time of writing; they can only be enacted once prior consent has been obtained.

12.9   Making Resident-Led Frailty-Prevention Services a Reality

In identifying risk factors, frailty prevention AI fully considers individual circumstances. However, close communication with residents is required when using AI outputs to guide service delivery. As discussed in Sect. 12.5, to deliver a bespoke set of frailty prevention services, we must understand the local service resources available for frailty prevention. We also need to consider who should deliver them and how they should be delivered to achieve a smooth transition to the delivery phase.

Therefore, as part of the Kashiwa-no-ha Smart-City project, we created a living lab on ideas for resident-led wellness services (this living lab is dubbed the "frailty-prevention AI living lab") in collaboration with the UDCK, which plays a key role in empowering residents to actively undertake wellness-promoting activities. See Chap. 7 (Citizen Participation) for more living lab initiatives.

With residents contributing their ideas, this living lab identified several ideas on how frailty prevention AI can be used. One idea is to link AI outputs with medical inspections so that residents, when receiving a medical inspection, can see their level of care-dependency risk and risk factors, along with recommended local services to address these risks. We are consulting Kashiwa City officials to make this idea realistic.

12.10   Toward a Sustainable Society of Smart Aging

I will end this chapter by discussing the kind of society we hope to create by implementing frailty prevention AI.

We envisage a more sustainable smart-aging society. To that end, we have developed what we call a "healthspan-society spiral" showing who does what to build such a society. Figure 12.11 outlines this spiral.

A spiral comprises three core components: The first component concerns people and activities related to health-span-focused community building. This component encompasses the first three phases of the spiral. Once resident-led frailty prevention activities are more evidence-based (see Sect. 12.3), more older adults will be healthy

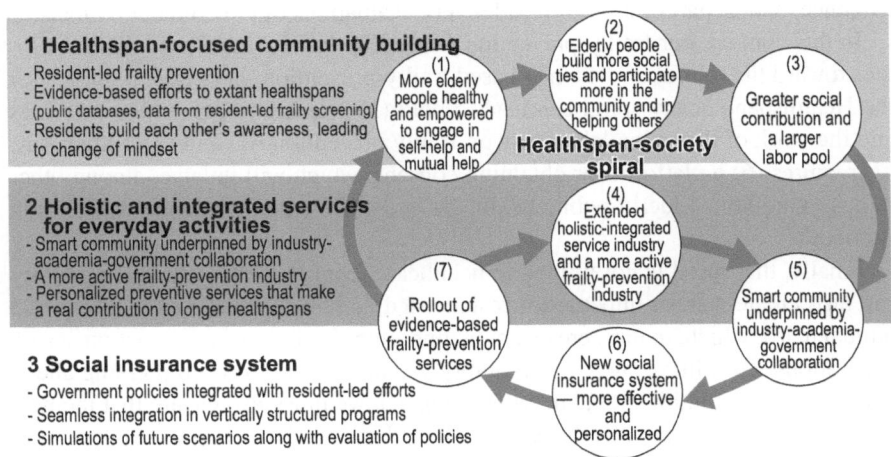

Fig. 12.11 Healthspan-society spiral

and empowered to engage in self-help and mutual help (1). This outcome encourages older adults to build social ties, participate more in the community, and help others (2). Outcome (2) results in greater social contributions and a larger labor pool (3).

The second component focused on the people and activities involved in holistic and integrated services for everyday activities. When frailty prevention services are based on solid evidence, they always lead to changes in everyday behavior. Such changes would suggest that there is an active frailty-prevention industry generating new value (i.e., delivering more effective support for preventing frailty). As this industry produces personalized support (as opposed to mass-producing one-size-fits-all goods and services), it is an extended holistic-integrated service industry and a more active frailty prevention industry (4).

The third component involves people and activities related to social insurance. Once the community has (3) (under the first component) and (4) (under the second component) in place, it can start working with people in social insurance to create a smart community underpinned by industry-academia-government collaboration (5). That is, when residents' health data and logged data about their everyday activities are collated, the data can be used by the public sector to develop new public services, and by the private sector to develop commercial goods and services. These services will spur greater self-help and mutual help in the community, leading to an innovative community in which digitized data are channeled to improve citizens' quality of life. As part of this process, multi-stakeholders confer on how to leverage digitized data to benefit the community. Such discussions will contribute to a new social insurance system that is more effective and personalized (Phase 6). Two things are of crucial importance here: connecting government strategies with resident-led efforts (a goal highlighted in a government program for integrating elderly public health services with care-dependency prevention services) and seamlessly linking vertically structured programs (wherein medical care is separated from nursing care). It is also helpful to run simulations of future scenarios based on evidence, and as part of a set of postdelivery evaluations of government policies.

In this context, our frailty-prevention AI corresponds to an activity that contributes toward the rollout of evidence-based frailty-prevention services (7). Phase 7 of the Healthspan-society spiral can lead to both (1) (via frailty prevention activities and the work of the living laboratory on frailty prevention AI) and (4). Either way, (7) requires (6) a platform for obtaining consent (which will involve, among other things, integrating elderly public health services with care dependency prevention services).

Finally, this spiral is not limited to healthcare. Digitizing the whole community can also create a fresh impetus for resident-led activities (involving self-help and mutual help), which, in turn, creates more opportunities for social interaction across generations. To achieve such a community vision, we need to keep pressing ahead with community digitization, but while doing so, we must also ensure that the new evidences and academic insights gleaned from big data pertaining to residents as a whole are effectively incorporated into their everyday lives. To this end, we must build social acceptance in the entire community by raising residents' awareness, fostering a change in mindset, and convincing residents that this is the best way forward. We must also ensure data governance in the handling of personal data contained within residents' health data, which residents are most concerned about. Such data governance requires the establishment of the necessary infrastructure. Alongside this, we must train talent to expedite efforts to strike a harmonious balance between building a warmhearted community and integrating the latest digital innovations. By integrating these components and the different architectures, we can make progress toward our vision of smart aging that is human-centric, smart aging that is expedited by citizen participation, and smart aging that is for the entire multi-generational community. Thus, this endeavor is not just about leveraging technology for its own sake; our frailty prevention AI should be delivered in the context of sustainable community development. As such, this challenging endeavor is a vital part of building innovative smart cities that embody "Society 5.0".

References

Cabinet Office (2022) Annual Report on the Ageing Society [Summary] FY2022. https://www8.cao.go.jp/kourei/english/annualreport/2022/pdf/2022.pdf. Accessed April 24, 2024. https://www8.cao.go.jp/kourei/whitepaper/w-2022/html/zenbun/index.html. Accessed April 24, 2024.

Cabinet Secretariat of Japan, Cabinet Office, Ministry of Finance, Ministry of Health, Labour and Welfare (2018) *2040 nen o misueta shakai hoshoo no shōrai mitōshi (giron no sozai)* [Social Insurance Projections for 2040: Resource for Discourse]. https://www.mhlw.go.jp/stf/seisakunitsuite/bunya/0000207382.html. Accessed April 24, 2024.

Fried, L. P., Tangen, C. M., Walston, J., Newman, A. B., Hirsch, C., Gottdiener, J., Seeman, T., Tracy, R., Kop, W. J., Burke and McBurnie, M. A. (2001) Frailty in Older Adults: Evidence for a Phenotype, The Journals of Gerontology: Series A, Volume 56, Issue 3, 1 March 2001, Pages M146–M157, doi:https://doi.org/10.1093/gerona/56.3.M146. Accessed October 18, 2024.

Hitachi (2019) Biomarker Search Service. https://www.hitachi.co.jp/products/it/industry/solution/hdsf_pharma/solution.html#a. Accessed April 24, 2022.

The Japan Geriatrics Society (2014) *Fureiru ni kansuru nihonrōnenigakukai kara no suteētomento* [The Japan Geriatrics Society's Statement on Frailty]. https://www.jpn-geriat-soc.or.jp/info/topics/pdf/20140513_01_01.pdf. Accessed April 24, 2024.

Kashiwa-no-ha Smart City Consortium (2021) *Kashiwa-no-ha sumātoshiti jikkō keikaku* [Kashiwa-no-ha Smart City Action Plan]. https://www.kashiwanoha-smartcity.com/action-plan/. Accessed April 24, 2024.

Kuzuya M (2009) *Rōnenigaku ni okeru sarcopenia & frailty no jūyōsei* [The Gerontological Importance of Sarcopenia and Frailty]. Journal of Japan Geriatrics Society 46(4), pp. 279–285. July 2009.

Lyu, W., Tanaka, T., Son, BK., Akishita, M. & Iijima, K. (2022) Associations of Multi-faceted Factors and Their Combinations with Frailty in Japanese Community-Dwelling Older Adults: Kashiwa Cohort Study, Archives of Gerontology and Geriatrics, Vol. 102, 104734, September-October 2022, DOI:https://doi.org/10.1016/j.archger.2022.104734.

Ministry of Health, Labour and Welfare (2021) *Shippei, shōgai oyobi shiin no tōkei bunrui* [Statistical Categories for Disease, Injury, and Cause of Death]. https://www.mhlw.go.jp/toukei/sippei/. Accessed April 24, 2024.

Ministry of Health, Labour and Welfare (2022) 2022 Edition Annual Health, Labour and Welfare Report: Maintain Sufficient Manpower to Support Social Security System, Part 1 (Outline). https://www.mhlw.go.jp/english/wp/wp-hw2022/dl/summary.pdf. Accessed April 24, 2024. https://www.mhlw.go.jp/stf/wp/hakusyo/kousei/21/index.html. Accessed April 24, 2024.

Tanaka, T., Son, BK., Lyu, W. & Iijima, K. (2022) Impact of Social Engagement on the Development of Sarcopenia among Community-Dwelling Older Adults: A Kashiwa Cohort Study, Geriatrics & Gerontology International, Vol. 22(5), pp. 384-391, March 2022, DOI:https://doi.org/10.1111/ggi.14372

Chapter 13
Project 3: City Infrastructure Management for Value Creation

Naoki Yoshimoto and Izuru Makihara

Abstract We proposed value creation that takes into account changes and issues in cities and regions in terms of infrastructure maintenance and management. There are three types of infrastructure value: basic value, sustainable value, and extended value. We analyzed and evaluated the characteristics that encompass these three types of value. We also used a causal loop diagram to analyze the factors that link regional characteristics to infrastructure maintenance and management. These value creation and evaluation methods make it possible to implement maintenance and management measures that respond to changes in the national land.

Keywords Infrastructure management · Value creation · Sustainable infrastructure · Causal loop diagram · City planning

13.1 Today's Infrastructure Management and Its Problems

This chapter outlines an approach for ensuring that in the process of building Society 5.0, the urban infrastructure and community assets (such as public facilities), which require a long service life, will be both accessible to residents and capable of generating new value through the use of digital technology.

Traditionally, before constructing urban infrastructure such as roads, bridges, sewer systems, or parks, a quantitative cost-benefit analysis is conducted. For roads or bridges, this typically involves calculating a benefit-cost ratio (B/C ratio), where

N. Yoshimoto (✉)
The Hitachi-UTokyo Laboratory, Research & Development Group, Hitachi, Ltd.,
Tokyo, Japan
e-mail: naoki.yoshimoto.rr@hitachi.com

I. Makihara
Political and Public Administration Systems, The Research Center for Advanced Science and Technology, The University of Tokyo, Tokyo, Japan
e-mail: makihara@pha.rcast.u-tokyo.ac.jp

© The Author(s) 2025
Hitachi-UTokyo Laboratory (H-UTokyo Lab.), *The Architecture of "Society 5.0"*,
https://doi.org/10.1007/978-981-96-2929-9_13

the benefit (B) represents the quantitative advantage of constructing the facility (e.g., the reduction in distance and time required to reach a destination), and the cost (C) includes the expenses involved in construction. For parks, benefits might be measured in terms of the percentage of green space, which indicates the proportion of area covered by trees and other vegetation. This conventional approach, based on the premise that urban infrastructure demands substantial investment and needs to be durable, has historically been the primary strategy.

Much of Japan's infrastructure was constructed during its period of high economic growth from the mid-1950s to the early 1970s. Currently, some of this infrastructure is now over half a century old and has become increasingly dilapidated. Consequently, inspection and repair work have generated significant public costs. However, infrastructure management has advanced significantly in technological innovation aimed at minimizing these costs. For example, inspections that once depended primarily on manual observation are now increasingly performed using drones, and Light Detection and Ranging (Lidar) technology is employed to map the undersides of bridges (Institute of Infrastructure Regeneration 2019). While this chapter does not explore individual instances of infrastructure management, it discusses how leveraging these technological advancements can enhance the value of infrastructure.

Historically, infrastructure management emphasized prospective evaluation, where a cost-benefit analysis was conducted with all estimates laid out before consulting an expert panel. This approach prioritized assessing whether the project would deliver good value for public funds by minimizing costs while ensuring the infrastructure remained functional and durable for many years. Initial designs become crucial, especially in large-scale projects where repair costs are likely to increase. Recently, however, the evaluation criteria have expanded to include community needs alongside the core functionality of the infrastructure. This shift reflects a broader understanding of how infrastructure projects should align with community issues and contribute to community development plans.

The government has taken a keen interest in infrastructure that drives community development and has launched a social-infrastructure funding program, formally known as the "comprehensive grant for developing social infrastructure." This program aims to provide flexible management of infrastructure projects (Ministry of Land, Infrastructure, Transport and Tourism 2024). It has two main components: First, it funds inclusive infrastructure projects that address local issues and needs effectively. Second, it supports projects that are not only inclusive but also flexible, enabling them to develop community-needs-focused social infrastructure. These projects are designed to envision a community-specific vision, establish Key Performance Indicators (KPIs) aligned with this vision, and utilize a retrospective evaluation process to assess their success.

13.2   What Is Infrastructure Management?

In this book, "infrastructure management" encompasses more than merely maintaining built infrastructure. It refers to a comprehensive approach that considers the long-term changes and needs of a city and its community, combining maintenance and renewal efforts to address these evolving demands. Historically, infrastructure maintenance was included in the pre-project cost-benefit analysis, which accounted for the expenses associated with regular upkeep and significant overhauls required as infrastructure neared the end of its useful life. However, this conventional approach has shown limitations. Often, the pre-project cost-benefit analysis does not account for reduced tax revenue from population decline, nor does it adjust benefits based on decreased utilization of the infrastructure. This oversight can lead to a dilemma: whether to invest in major improvements or to completely replace the infrastructure. Furthermore, with societal shifts—such as population decline, changes in transport and mobility, and lifestyle alterations—occurring more rapidly than the designed lifespan of the infrastructure, there is a pressing need for the infrastructure to adapt promptly to these changes.

This book identifies six key factors essential for building a smart city, among which infrastructure management is pivotal. Infrastructure management acts as an interface for social infrastructure in cities and communities, linking the social infrastructure with the people and smart city architecture through two key factors: citizen participation and smart-city Quality of Life (QoL) assessments. Assuming the following examples of infrastructure management projects, data from QoL assessments and analyses of data generated from citizen participation are incorporated into infrastructure management.

This integration helps in understanding what community needs and issues have been considered in managing infrastructure. Figure 13.1 illustrates the community KPIs linked to the government's Comprehensive grant for developing social infrastructure. These KPIs are divided into two broad categories: KPIs that promote economic activities beneficial to the local economy and residents' lives and KPIs aimed at creating vibrant urban centers and improving transportation links, including initiatives such as redeveloping station vicinities and widening roads (Institute for Future Urban Development 2022).

Figure 13.1 presents examples of the KPIs utilized in such infrastructure projects. The selection of KPIs includes factors such as the target area's population, transport volume, number of shops, and land prices. These indicators highlight how local infrastructure projects are progressively incorporating community themes into their designs. Unlike in the past, where projects were primarily evaluated based on the infrastructure's functionality, current projects also use these themes as KPIs to assess their effectiveness.

These projects aim to address community needs and issues through tailored infrastructure management. While local conditions are considered in these initiatives, they are still in the preliminary stages of developing a comprehensive framework for managing the infrastructure. This includes defining the overarching vision

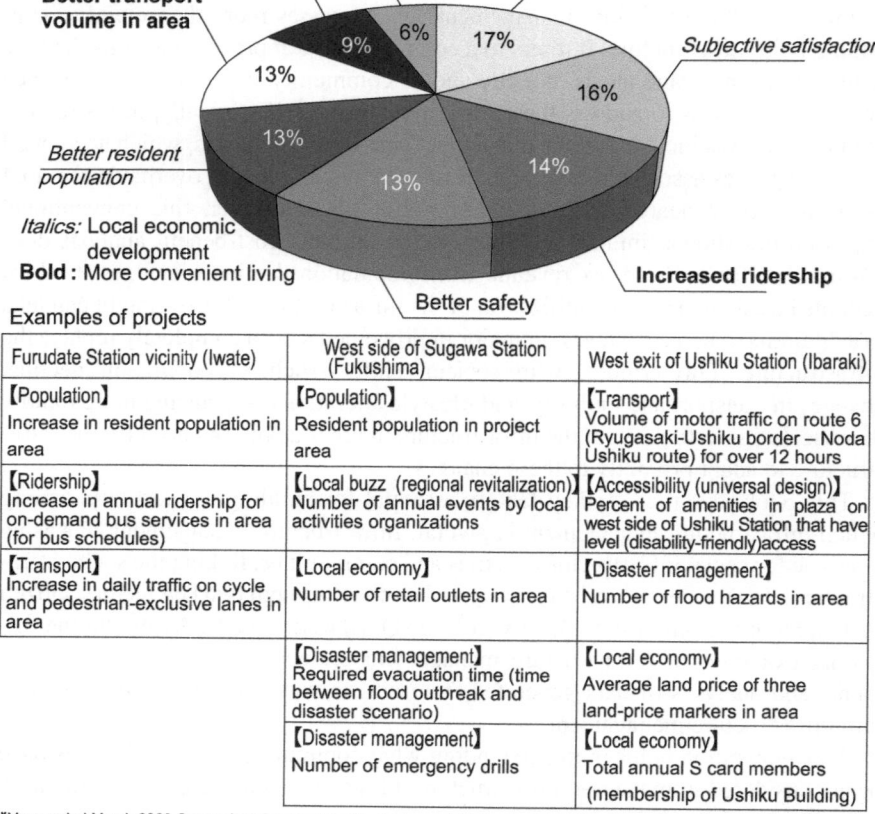

Fig. 13.1 Social-infrastructure grant breakdown

Furudate Station vicinity (Iwate)	West side of Sugawa Station (Fukushima)	West exit of Ushiku Station (Ibaraki)
【Population】 Increase in resident population in area	【Population】 Resident population in project area	【Transport】 Volume of motor traffic on route 6 (Ryugasaki-Ushiku border – Noda Ushiku route) for over 12 hours
【Ridership】 Increase in annual ridership for on-demand bus services in area (for bus schedules)	【Local buzz (regional revitalization)】 Number of annual events by local activities organizations	【Accessibility (universal design)】 Percent of amenities in plaza on west side of Ushiku Station that have level (disability-friendly)access
【Transport】 Increase in daily traffic on cycle and pedestrian-exclusive lanes in area	【Local economy】 Number of retail outlets in area	【Disaster management】 Number of flood hazards in area
	【Disaster management】 Required evacuation time (time between flood outbreak and disaster scenario)	【Local economy】 Average land price of three land-price markers in area
	【Disaster management】 Number of emergency drills	【Local economy】 Total annual S card members (membership of Ushiku Building)

*Year ended March 2020 Comprehensive grant for developing social infrastructure: City regeneration, city rebuilding

of the project. It is crucial to specify what values the residents seek, how these values can be delivered, and how to establish sustainable infrastructure management that will enhance both the community and the residents' QoL.

13.3 Three Types of Value Infrastructure Delivers

When establishing goals and KPIs for infrastructure, it is essential to concentrate on the values important to residents: the conventional performance of existing infrastructure, residents' satisfaction, economic development, and community revitalization. We have identified three categories of value for residents, which are aligned with the components (≒ value items) of QoL (Fig. 13.2). The first is basic value,

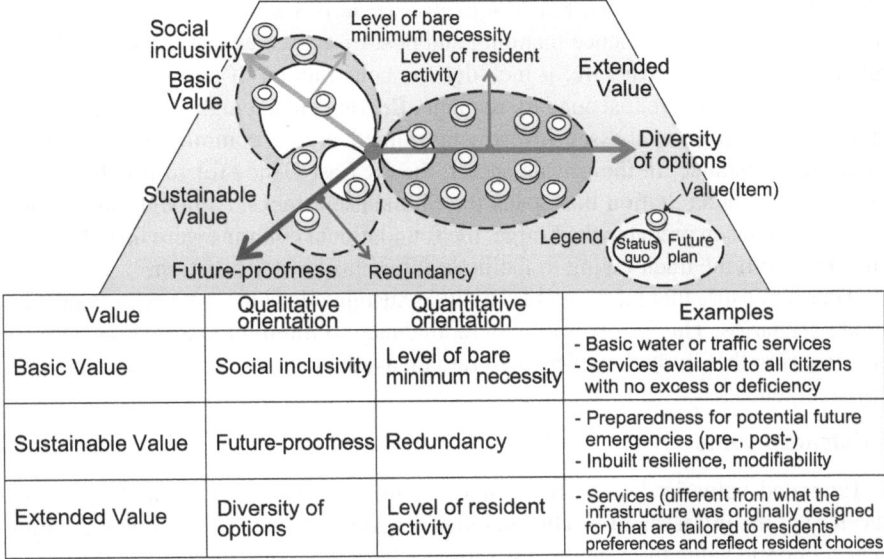

Fig. 13.2 Value categories based on QoL elements in infrastructure management

Value	Qualitative orientation	Quantitative orientation	Examples
Basic Value	Social inclusivity	Level of bare minimum necessity	- Basic water or traffic services - Services available to all citizens with no excess or deficiency
Sustainable Value	Future-proofness	Redundancy	- Preparedness for potential future emergencies (pre-, post-) - Inbuilt resilience, modifiability
Extended Value	Diversity of options	Level of resident activity	- Services (different from what the infrastructure was originally designed for) that are tailored to residents' preferences and reflect resident choices

reflecting the infrastructure's original design purpose. The second is sustainable value, focusing on the role of infrastructure in enhancing community preparedness for natural disasters or future developments. The third is extended value, relating to the infrastructure's ability to serve additional purposes beyond its original design.

1 Basic value

Basic value represents the essential services that infrastructure must provide, and its presence is typically taken for granted; however, its absence leads to severe dysfunction. For water infrastructure, the basic value is realized through the consistent delivery of safe, high-quality water and a reliable water supply. In terms of traffic infrastructure, it manifests as the provision of well-maintained roads and efficient mobility.

In many cases, basic value signifies the delivery of a universal service—something that should be consistently available across the community to ensure social inclusion now and in the future. Therefore, to maintain the infrastructure's basic value, it is crucial not to pursue economic efficiency to such an extent that it renders the service inaccessible to some residents. Efforts should be made to eliminate waste resulting from over-supplying the service, and it is important to minimize any excess or deficiency by delivering the service in a manner that deviates from the traditional infrastructural assets.

2 Sustainable value

Sustainable value might not seem critical at present, but it becomes essential in future scenarios or emergencies, where it is indispensable for the continuity and resilience of the community. For water infrastructure, sustainable value is realized through earthquake-resilience planning and post-disaster recovery strategies. In the realm of traffic infrastructure, it includes the construction of anti-flood reservoirs and initiatives to prevent secondary damage. Beyond enhancing resilience, sustainable value also encompasses investments aimed at the community's long-term future. For instance, in the private sector, the JR East Ueno–Tokyo line features a bridge near Kanda Station that spans the Shinkansen tracks. Although this section of the Shinkansen predates the bridge, the foundational columns were installed concurrently with the track laying to facilitate future bridge construction.

Thus, to ensure that infrastructure delivers sustainable value, we need to consider future scenarios. This involves adding redundancy to future infrastructural services and ensuring that limited services will remain available to vulnerable residents during emergencies.

3 Extended value

Extended value is derived from functionalities beyond the originally intended services of the infrastructure. This includes generating a buzz in the area or creating new economic value through the utilization of both the infrastructure and its related data. For example, in water infrastructure, extended value can arise when the delivered water is notably tasty or when facilities such as water treatment plants are made accessible to the public, thereby garnering affection and support for the installations. In the realm of traffic infrastructure, extended value can be realized when roadways offer scenic views, such as vistas of private gardens along the route.

Extended value can be categorized into three distinct types. The first category is upgrading, where the extended value is derived from services that fall within the infrastructure's typical usage scope, enhancing convenience, efficiency, and comfort for users. The second category, diversification, originates from services outside the normal scope of the infrastructure's use, which contribute to emotional well-being (such as services that offer healing or promote a caring state of mind). The third category, external extended value, pertains to economic value generated by services that utilize another piece of infrastructure alongside the original, or use data associated with the original infrastructure.

Extended value often requires tailoring services to residents' preferences and providing diverse options. Additionally, it is crucial that this value supports local economic development by facilitating an increase in residents' activity levels.

An illustrative example of generating extended value is the Paris-Plages (see Fig. 13.3). Paris-Plages transforms sections of roadway along the Seine in central Paris into artificial beaches during the summer. Typically, many Parisians prefer to leave the city during the hot summer months; however, the increasing tourist influx necessitates that many remain to work in the service industry. To enhance the QoL for these residents, the mayor initiated a plan to close the riverside roadways to traffic during the summer (when commuter traffic is reduced) and convert them into public resorts complete with palm trees and sand. Originally designed for transportation, these roadways now also maximize the scenic value of the Seine, thereby significantly enhancing Parisians' QoL.

Paris-Plages: Public beaches on roadways along the Seine

- During the summer-holiday period, many Parisians prefer to visit the seaside or countryside to avoid the city's sweltering heat. However, due to the influx of tourists to Paris during this time, numerous residents find themselves staying in the city.
- To accommodate these Parisians, artificial beaches have been established along roadways adjacent to the Seine, where traffic diminishes in the summer-holiday period. Paris-Plages (initially named Paris-Plage) was inaugurated for public use in 2002.

Fig. 13.3 Example of extended value creation: Paris-Plages

13.4 Value-Creation Process

For infrastructure management to be truly resident-oriented, it must prioritize community needs and issues over mere maintenance and renewal. This approach necessitates engaging in specific processes that align with two of the six key factors identified for effective management: citizen participation and smart-city QoL assessment. Figure 13.4 illustrates the value-creation process essential for this approach in infrastructure management.

To ensure infrastructure management is effectively oriented toward community needs, the first step involves collecting data about the community's features. This data collection process should ideally utilize open data due to its accessibility. Subsequently, this data is analyzed to calculate the quantity of features, which assists in assessing community needs accurately. Given the current trends, government data are increasingly available as structured datasets. This format simplifies the use of statistical data, making it accessible even for those who are not typically adept at handling statistical information.

The subsequent process involves evaluating the quantities of community features. In this evaluation, community feature quantities are categorized and rated using a relative grading system to facilitate intuitive visualization. Figure 13.4 illustrates this evaluation using a radar chart, which provides a visual comparison between the community and another city of similar population size. Typically, municipal-level data are utilized to identify these features and perform comparisons

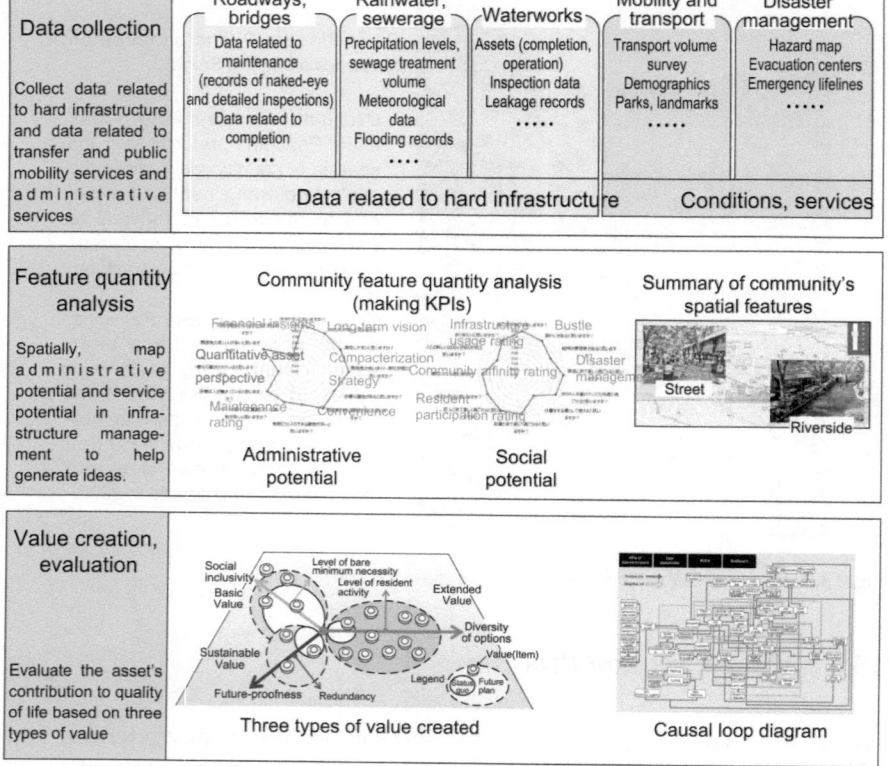

Fig. 13.4 Value-creation process in infrastructure management

with comparable cities. However, infrastructure management often focuses on a sub-municipal level, such as a community or district, where necessary data may not always be readily available. In instances of data mismatch between the collected statistics and the target community, it is advisable to consult the community needs/ issues and the KPIs outlined in the previously mentioned Comprehensive grant for developing social infrastructure.

The final step involves mapping the community needs and issues with infrastructure management to devise a plan that generates the three types of values: basic, sustainable, and extended. During this planning phase, it is crucial to incorporate KPIs to assess whether the project is truly resident-oriented and effectively addresses community needs and issues. The next section introduces a causal loop paradigm that connects infrastructure with a community's needs.

13.5   Causal-Analysis Technique for Creating Community Value

To effectively integrate community features with infrastructure management, it is essential to understand how community needs and issues are related to infrastructure capabilities. Unfortunately, there are few examples that illustrate how infrastructure can deliver significant value to the community. This scarcity is largely due to the traditional approach to infrastructure management, which primarily focuses on maintaining infrastructure in operational condition rather than enhancing its utility to meet community needs. Given this context, it may be beneficial to explore the relationship between infrastructure and resident services more deeply, especially considering that the social-infrastructure grant will now be utilized in infrastructure renewal projects. However, there is also a challenge in this area: Determining the relationship between infrastructure and resident services is complicated by the limited number of cases that involve large-scale renewal efforts or well-developed infrastructure project proposal.

Thus, identifying community features and integrating infrastructure management with the community's needs and issues is essential, though laborious. This integration necessitates a thorough understanding of the on-the-ground realities and a realistic operational plan for the infrastructure. To facilitate this complex process, one effective strategy is to develop a platform employing a causal-analysis technique. This platform will foster a shared perspective between the regional agency managing the infrastructure and the residents it serves.

Causal analysis helps to visualize a causal chain. To illustrate using a hypothetical example from the Japanese proverb, "When the wind blows, bucket-makers prosper"—akin to the butterfly effect where a minor incident triggers a significant event—causal analysis elucidates how seemingly unrelated events are connected. The causal-analysis technique discussed in this context is the causal loop diagram (CLD), which graphically represents the factors linking the wind's impact to the prosperity of bucket-makers. Figure 13.5 displays a CLD that outlines the programs and infrastructure causally linked to residents' QoL. This diagram highlights the extensive range of factors that influence QoL, including traffic conditions, environmental quality, the demand for green spaces, and lifestyle preferences, such as the propensity for walking. In this chapter, "QoL" is used in a broader sense to include, in addition to QoL, traffic, environment, and lifestyle preferences. To enhance QoL, it is recommended to use tools that consider an individual's activities. One such tool, Active QoL, is discussed in this book. The CLD shown in the figure can illustrate the KPIs relevant to each user and depict the connections between these KPIs throughout the plan-do-check-act cycle. This intuitive visual aid offers a shared reference point for stakeholders, such as representatives of local public organizations and residents, who may approach the topic from diverse viewpoints. This makes it a valuable method for engaging with nonexperts.

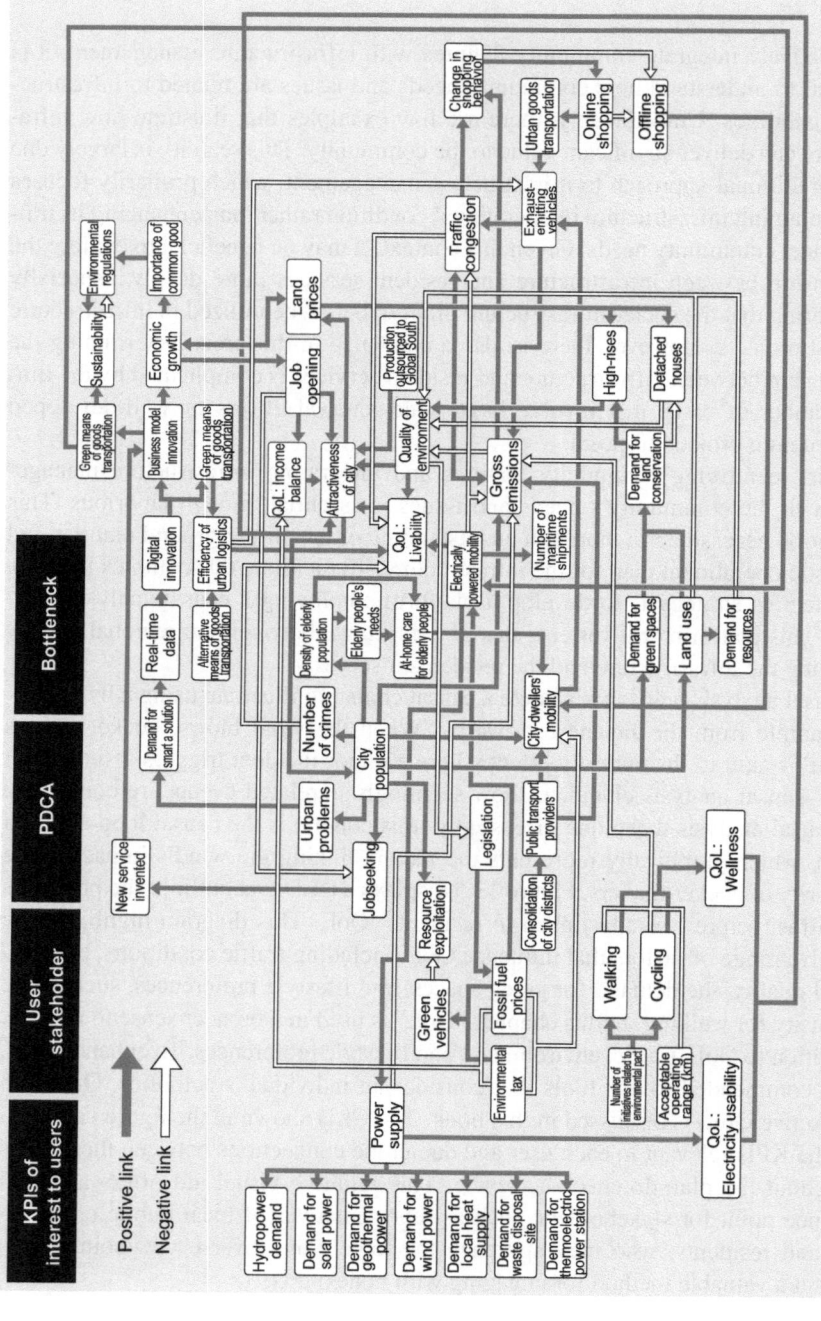

Fig. 13.5 Example of a causal loop diagram

Constructing a CLD from scratch is a challenging process, as it requires the meticulous identification and integration of all causal factors involved. The availability of templates—illustrative examples of CLDs that outline potential factors in a community setting—significantly aids this process. When these templates are digitally accessible, allowing for intuitive navigation (e.g., by tapping on a tablet), they can be readily used and tailored to fit specific community scenarios. With an increasing number of causal analyses being conducted across various communities, it becomes feasible to evaluate the impact of projects comprehensively and to initiate new programs accordingly. Such initiatives enhance the potential for observing the ripple effects within a single CLD. Although creating a CLD from the ground up is formidable, utilizing an existing template from another community and adapting it to incorporate the unique issues and needs of the current community can substantially mitigate this challenge.

The final point to note is that processing causal factors and making future projections can be achieved with greater precision if each correlation is mathematically formulated (expressed as a function). The adoption of digitization will aid in eliminating interregional disparities, such as variations in funding among regional agencies, and will facilitate mutual learning among communities based on each other's programs.

13.6   Toward Infrastructure Management That Creates Value

In this chapter, the concept of infrastructure management aimed at creating value is outlined, with various examples cited to illustrate this approach. Japanese rural areas are experiencing population declines faster than initially anticipated, indicating that patterns of sparsely populated communities are likely to dominate much of Japan in the long term. This shift in land use, along with the challenges of managing infrastructure, should ideally serve as an impetus for both residents and the government to collaboratively consider the long-term prospects for needs-based infrastructure management.

References

Institute for Future Urban Development (IFUD) (2022) *Machizukuri jōhō kōryū shisutemu* [Community-building information exchange system]. http://www.machikou-net.org/. Accessed April 30, 2024.

Institute of Infrastructure Regenerationfura, Nikkei Construction (ed) (2019) *Kōhaisuru nihon kore de ii no ka japan infura* [Is this what you want, Japan Infrastructure?]. Tokyo, Nikkei Business Publications

Ministry of Land, Infrastructure, Transport and Tourism (2024) *Shakai shihon seibi sōgō kōfu kin tō ni tsuite* [On the comprehensive grant for developing social infrastructure]. https://www.mlit.go.jp/page/kanbo05_hy_000213.html. Accessed April 30, 2024.

Chapter 14
Six Key Factors as Infrastructure for a Digital Society

Tomoyo Sasao, Shin Osaki, Hideyuki Matsuoka, Soichi Furuya, and Atsushi Deguchi

Abstract This chapter introduces the six key factors described so far as infrastructure for the digital society and the tools necessary to utilize them toward creating a people-centric sustainable smart city. In order for a wide variety of players participating in smart cities to collaborate under a common goal, one of the challenges is to have a common understanding of how far each city and area has been formed and what the next goal is. Besides, there is no common rubric that clearly articulates the expectations for each developmental step or level toward achieving a people-centric city. This chapter discusses in the following structure.

Section 14.1 summarizes such challenges to consensus building and phased development. Section 14.2 defines the steps to incrementally implement six key factors in cities and regions. Here a common four-step template is given, which is then applied to six key factors, i.e., Social Acceptance, Data Governance, Citizen Participation, QoL-based Assessment, Human Resource Development, and Data Ecosystem. Section 14.3 discusses the phased implementation of smart cities using

T. Sasao (✉)
Faculty of Engineering, Reitaku University, Chiba, Japan
e-mail: tsasao@reitaku-u.ac.jp

S. Osaki
Neighverse Inc., Tokyo, Japan

Sustainable Society Design Center, Graduate School of Frontier Sciences, The University of Tokyo, Tokyo, Japan
e-mail: osaki@edu.k.u-tokyo.ac.jp

H. Matsuoka
The Basic Research Center, Research & Development Group, Hitachi, Ltd., Tokyo, Japan
e-mail: hideyuki.matsuoka.ws@hitachi.com

S. Furuya
AI Transformation Strategy, Digital Engineering Business Unit, Hitachi, Ltd., Tokyo, Japan
e-mail: soichi.furuya.xz@hitachi.com

A. Deguchi
Department of Socio-Cultural Environmental Studies, Graduate School of Frontier Sciences, The University of Tokyo, Tokyo, Japan
e-mail: deguchi@edu.k.u-tokyo.ac.jp

Hitachi-UTokyo Laboratory (H-UTokyo Lab.), *The Architecture of "Society 5.0"*,
https://doi.org/10.1007/978-981-96-2929-9_14

these concepts, and finally Sect. 14.4 summarizes the key factors as infrastructure in a digital society.

Keywords Maturity level of smart cities · Six key factors for a people-centric sustainable smart city · Incremental actions · Framework · Assessment

14.1   Shared Frame of Reference for Creating People-Centric Smart Cities

Smart cities represent an attempt to create people-centric cities through ambitious programs designed to address community needs and enhance residents' QoL. The success of these initiatives hinges on the collaborative efforts of all parties involved in the project, each contributing unique insights and resources. This collaboration, though critical, is not straightforward. Given that these stakeholders represent a wide range of industries and sectors, defining the progressive steps toward establishing a people-centric city is challenging. It is particularly difficult to develop a shared understanding of these steps, which is essential for applying them to the specific actions undertaken by different actors.

At the H-UTokyo Lab, we recognized the potential for gaining insights into the integration of vehicle automation systems, which have advanced significantly by utilizing a variety of underlying technologies for complex processing tasks. We explored methods to apply the frameworks used in these technological advancements to smart cities.

In the context of vehicle automation systems, the "framework" offers a standardized language that articulates the future vision of vehicle automation along with a structured progression toward this vision through defined levels of automation (refer to Fig. 14.1). Initially proposed by the SAE in 2014 (SAE International 2021), these levels have undergone several revisions but fundamentally have remained consistent. They delineate five levels of automation (0–5), with higher levels indicating greater automation. Each level defines a common goal and a set of conditions,

Level	Name	Vehicle driven by	Restrictions on driving
Level 0	No driving automation	Human driver	–
Level 1	No driving automation	Human driver	Restricted
Level 2	Partial driving automation	Human driver	Restricted
Level 3	Conditional driving automation	Vehicle	Restricted
Level 4	High driving automation	Vehicle	Restricted
Level 5	Full driving automation	Vehicle	Unrestricted

Fig. 14.1 SAE levels of automation. (Source: SAE International 2021)

facilitating smooth collaboration among developers from various fields by providing a shared understanding. This framework not only offers a global perspective but also serves as a foundational tool for international collaboration. For example, Japan has utilized these common definitions to create a detailed roadmap that specifies the target levels of automation to be achieved by specific years (Strategic Conference for the Advancement of Public and Private Sector Data Utilization 2021). By 2020, Japan aimed to commercialize vehicles with level 2 automation, which supports steering, braking, and acceleration, and enables hands-free driving on expressways. Additionally, Japan has initiated pilot projects for level 3 and level 4 automation in various regions. This approach to defining automation levels exemplifies how a clear, common framework can facilitate complex collaborations across different sectors, speeding up innovation by aligning various stakeholders on current objectives, upcoming challenges, and future milestones.

Let us consider how the framework used for vehicle automation can be applied to smart cities. In smart city projects, similar to vehicle automation systems, there is a necessity for high-level collaboration among actors from diverse fields, often including residents, who work together to develop services and technologies for the smart city. Thus, smart cities necessitate a coordinated approach as they involve complex system development akin to vehicle automation.

It is widely acknowledged that smart city projects require a shared frame of reference or rubric. For this purpose, indices to assess and guide smart city development have been proposed globally. For instance, in 2017, the International Organization for Standardization (ISO) published ISO 37153, which is a maturity model for assessment and improvement of smart community infrastructures. This standard evaluates the maturity of smart city infrastructure—encompassing energy, water, transportation, waste management, and ICT—across three aspects: technical performance, process, and interoperability.

The standard defines five maturity levels: initial, partially fulfilled, fulfilled, improving, and sustainably optimizing (International Organization for Standardization 2017). These levels are analogous to those found in models used in organizational management, such as the Plan-Do-Check-Act (PDCA)[1] cycle and the Capability Maturity Model Integration (CMMI),[2] which facilitate continuous improvement. Another pertinent example is ITU-T Y.4904/L.1604, which is the smart sustainable cities' maturity model. Linked to the Sustainable Development Goals (SDG), this model categorizes city sustainability into three maturity dimensions: social, economic, and environmental, each with five defined maturity levels (International Telecommunication Union 2020).

---

[1] The plan-do-check-act (PDCA) cycle is a cycle of hypothesis and testing for continuous improvement. It was first proposed in the 1950s by William Edwards Deming, hailed as the father of quality management.

[2] Capability maturity model integration (CMMI) is used in efforts to improve an organization and project processes. The CMMI defines five maturity levels for project processes and the organization: (1) initial, (2) managed, (3) defined, (4) quantitatively managed, and (5) optimizing.

Thus, we possess instruments capable of assessing the maturity levels of smart cities. However, these instruments do not yet incorporate the people-centric focus central to Society 5.0, which prioritizes the well-being of city residents. Consequently, there is no common rubric that clearly articulates the expectations for each developmental step or level toward achieving a people-centric city.

14.2 Defining the Levels of Applying the Six Key Factors

Previous chapters have emphasized the significance of six key factors essential for developing a people-centric smart city. This chapter delineates the steps or levels required for their application in urban and community settings. We introduce this framework as a new rubric intended to assist all stakeholders involved in the creation of a people-centric smart city. This rubric aims to foster a shared understanding of the project's current progress and its ultimate objectives. It is our hope that this framework will serve as a valuable tool in advancing the development of smart cities.

Figure 14.2 illustrates four progressive levels representing the stepwise application of the six key factors within the target community. At level 0 (nothing done), no actions have been taken with consideration of the key factors. At level 1 (research done), research has been conducted to explore methods for applying these factors within the target community. Level 2 (partially implemented, trialed) denotes partial implementation of the key factors or piloting of methods researched at level 1. At level 3 (implemented, standardized), a body of practical experience and best practices has accumulated, resulting in the integration of the key factors as fundamental mechanisms in smart-city operations, thereby ensuring effective functionality in

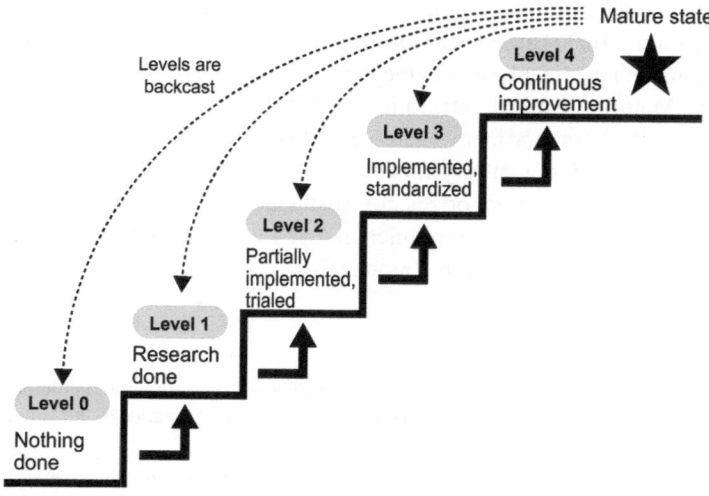

Fig. 14.2 Stepwise levels in the application of the key factors

practice. Level 4 (continuous improvement) describes the ideal scenario where the key factor functions effectively within the target community and involves an ongoing cycle of enhancements aimed at maintaining its effectiveness. Such improvements may involve closer alignment of the key factor with the community's realities or actions to enhance its stability.

Figure 14.3 utilizes the foundational model of levels and applies it to each key factor, projecting a progression from the ideal scenario toward specific levels tailored to each factor. These levels delineate the necessary steps to advance toward the ideal scenario. With levels 1 and 2 encompassing relatively minor initial steps, the model encourages an initial focus on reaching level 3. Trials are initiated with the aim of attaining this level.

The subsequent section of this chapter proposes a series of incremental actions to optimize the functionality of the six key factors within the target community. Additionally, it highlights essential perspectives to consider while undertaking these actions.

14.2.1 Key Factor 1: Social Acceptance

For the social acceptance factor, we envisaged the desirable endpoint—the public are well informed about and receptive to the smart city programs. We developed a range of levels, starting from this goal, to show different stages of improvement in public understanding and acceptance of smart city programs (see Fig. 14.3).

Level 0 signifies a lack of progress, where no actions have been taken. At this stage, the smart city operator lacks understanding of social acceptance, as discussed in Chap. 5.

Level 1 signifies a stage where research has been conducted to explore approaches for enhancing social acceptance. At this level, the smart city operator has gained an understanding of social acceptance, is engaged in researching approaches for fostering it, and is contemplating adoption strategies for the organization.

Level 2 represents the trial stage. At this level, programs for building social acceptance are being sporadically trialed.

Level 3 is the organizational embedding stage. At this level, decisions are being made based on the results of trials on initiatives for building social acceptance, which initiatives to adopt, and how to sustainably entrench these initiatives.

Level 4 represents the stage of continuous improvement. At this level, the programs and mechanisms integrated during level 3 are periodically refined, and a budget and framework are in place to enable continuous improvement.

The exact approach taken to building and sustaining social acceptance will depend on the community in question. One method may work for building rapport with citizens in a big city, and this will differ from one that would work in a rural or remote municipality. At level 1, an approach that works for the community in question needs to be identified. Insights into this can be gleaned by using a model representing the attitudinal structure of social acceptance toward a smart city. However,

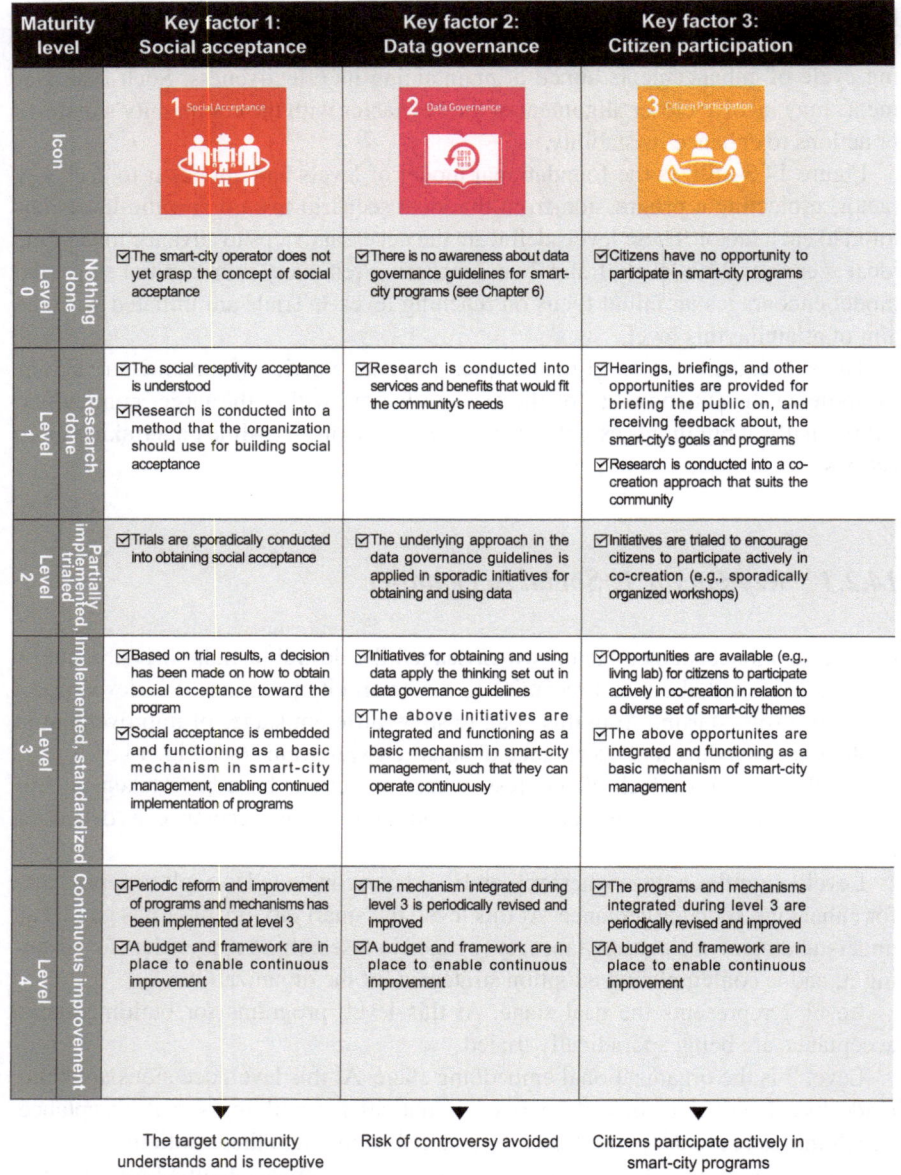

Maturity level		Key factor 1: Social acceptance	Key factor 2: Data governance	Key factor 3: Citizen participation
Icon		1 Social Acceptance	2 Data Governance	3 Citizen Participation
Level 0	Nothing done	☑The smart-city operator does not yet grasp the concept of social acceptance	☑There is no awareness about data governance guidelines for smart-city programs (see Chapter 6)	☑Citizens have no opportunity to participate in smart-city programs
Level 1	Research done	☑The social receptivity acceptance is understood ☑Research is conducted into a method that the organization should use for building social acceptance	☑Research is conducted into services and benefits that would fit the community's needs	☑Hearings, briefings, and other opportunities are provided for briefing the public on, and receiving feedback about, the smart-city's goals and programs ☑Research is conducted into a co-creation approach that suits the community
Level 2	Partially implemented, Implemented, trialed	☑Trials are sporadically conducted into obtaining social acceptance	☑The underlying approach in the data governance guidelines is applied in sporadic initiatives for obtaining and using data	☑Initiatives are trialed to encourage citizens to participate actively in co-creation (e.g., sporadically organized workshops)
Level 3	Implemented, standardized	☑Based on trial results, a decision has been made on how to obtain social acceptance toward the program ☑Social acceptance is embedded and functioning as a basic mechanism in smart-city management, enabling continued implementation of programs	☑Initiatives for obtaining and using data apply the thinking set out in data governance guidelines ☑The above initiatives are integrated and functioning as a basic mechanism in smart-city management, such that they can operate continuously	☑Opportunities are available (e.g., living lab) for citizens to participate actively in co-creation in relation to a diverse set of smart-city themes ☑The above opportunites are integrated and functioning as a basic mechanism of smart-city management
Level 4	Continuous improvement	☑Periodic reform and improvement of programs and mechanisms has been implemented at level 3 ☑A budget and framework are in place to enable continuous improvement	☑The mechanism integrated during level 3 is periodically revised and improved ☑A budget and framework are in place to enable continuous improvement	☑The programs and mechanisms integrated during level 3 are periodically revised and improved ☑A budget and framework are in place to enable continuous improvement
		▼ The target community understands and is receptive toward smart-city programs	▼ Risk of controversy avoided	▼ Citizens participate actively in smart-city programs

Fig. 14.3 Levels (and their criteria) in the implementation of each of the six key factors

Key factor 4: QoL-based assessment	Key factor 5: Human resource development	Key factor 6: Data ecosystem
4 QoL-Based Assessment	**5** Human Resource Development	**6** Data Ecosystem
☑There is no (or there is no awareness about) a means for checking whether a smart-city programs is contributing to QoL	☑There is no conscious attempt to develop a human-resource-development strategy for the running of the smart city (e.g., staff left to fend for themselves)	☑There is no inter-organizational data sharing in smart-city programs
☑Research is conducted into a technique for measuring residents' subjective metrics ☑Metrics suited to community have been selected	☑Research is conducted into how to standardize human resource development	☑Research is conducted into the needs and seed capital related to inter-organizational data use
☑Smart-city programs are sporadically evaluated using subjective metrics (e.g., QoL)	☑Ad hoc human-resource-development programs are trialed sporadically (e.g., external training scheme, internal training system)	☑Inter-organizational data use is trialed sporadically in relation to a particular smart-city theme
☑Such evaluations are integrated and functioning as a basic mechanism of smart-city management, with a PDCA cycle conducted to identify community issues and rectify them	☑There is a clear vision setting out the smart-city human resources that the target community requires ☑Human-resource-development programs aligned with the above vision are selected and implemented ☑The programs are integrated and functioning as a basic mechanism of smart-city management, enabling continuous human resource development	☑Services involving inter-organizational data use are in operation and are delivering the expected benefits and revenue ☑Effective strategies for supporting data use are derived from high-quality use cases ☑The support strategies are integrated and functioning as basic mechanism in smart-city management
☑The evaluation process integrated during level 3 is periodically reformed and improved ☑A budget and framework are in place to enable continuous improvement	☑The programs integrated during level 3 are periodically reformed and improved ☑A budget and framework are in place to enable continuous improvement	☑The services based on data linkage are reformed and improved—e.g., feedback from residents and other stakeholders are taken into account and services are discontinued, new services are added, or services are adapted to changing needs ☑Periodic reform and improvement of support programs ☑A budget and framework are in place to enable continuous improvement
▼	▼	▼
Data are used to show how smart-city programs are contributing to QoL	A workforce has been cultivated for managing the smart city	Inter-organizational data use increases based on resident needs, contributing to city sustainability

Fig. 14.3 (continued)

the general principle is that trust in the smart-city operators is a critical determinant of social acceptance. Trust itself is shaped by procedural fairness (whether the processes adopted are fair) and distributive fairness (whether the benefits are distributed proportionately, as opposed to disproportionately benefiting a particular community, gender, or generation, for example). Thus, regardless of the kind of community in question, at level 3 procedural and distributive fairness must be kept in mind when trying to engage residents in dialog.

The general approach to ensuring procedural fairness is to only proceed with the project after a robust dialog has been held with the right stakeholders and consent obtained from all the stakeholders. Who these stakeholders should be and when it is best to hold the dialog with them will depend on the community and project in question. A good way to determine these matters is to have a strong understanding of the community's interests and to have a plan for reaching out effectively to the stakeholders. To that end, a wide range of stakeholders should be considered—not just local such as town meetings and the local chamber of commerce—and the dialog approach should be made as inclusive as possible.

As for distributive fairness, you must try to avoid any disproportionate allocation of benefits. Oftentimes, however, when one becomes obsessed with trying to achieve distributive fairness in a single project, the project target becomes obscured, and the project goes awry. It is better, then, to group projects together and try to ensure distributive fairness for the project group as a whole. You should also try to ensure equality in dialog opportunities. Given the intergenerational gap in digital literacy, it might be a good idea to provide elderly people with in-person briefings and introduce the jargon in a digestible format. For younger people who are more digitally literate, you can use an online approach to collect feedback.

14.2.2 Key Factor 2: Data Governance

It is now a global trend to take a data-driven approach to address community needs and issues and improve residents' QoL. In data-driven smart cities, there are platforms linking data between different fields to enable solutions for complex problems and creating potential for new business opportunities to emerge. However, to avoid trouble and conflict in such data use, we need robust data governance, which includes, in addition to legal and regulatory compliance, measures for preventing controversy and scandal.

Chapter 6 presented basic data governance guidelines for smart cities. The maturity levels for data governance represent the degree to which such guidelines are being applied (Fig. 14.3).

Level 0 represents a state in which initiatives are being conducted with no awareness about data governance guidelines. At this level, no thought is given to the need for legal and regulatory compliance or the need to avoid controversy or scandal. As such, trouble and conflict are highly likely to be encountered.

At level 1, services and benefits that could be suitable for the community have been investigated, referring to data governance guidelines. Data use is a means, not an end, so a start must be made by involving residents and working out what kind of value should be delivered to residents.

Level 2 represents the stage where a clearly defined purpose for using the data considered during level 1 exists and where the data has started to be used. Following the guidelines, we sporadically obtain and use data that is necessary for implementing new services. It is a good idea to progressively rectify and refine underway data-driven programs based on the guidelines.

Level 3 represents a stage where data-driven initiatives have been delivered in the community based on the guidelines and where such data use is now integrated and functioning as a basic mechanism in the management of the smart city. At this level, the guidelines-based creation and outcomes of services are evaluated, and the results of evaluations are used to guide forthcoming programs.

Level 4 represents a stage where the actions of level 3 are practiced and improved in a sustained cycle. Agile governance (see Sect. 6.3) is useful at this level. A system of data governance is an unfinished product from the start; it is best to operate a smooth cycle of trial and improvement. At this level, the consent of the stakeholders and effective budget allocations have been achieved.

At this level, the goal is to have a smooth, conflict-free process of using data in the smart city. To that end, the data governance guidelines must be generic to some extent. The outcomes of applying these guidelines in practice should then be considered so that the guidelines can be refined and rolled out in different communities. In this way, the guidelines will become increasingly relevant to the situation on the ground and increasingly usable, reaching a stage where the guidelines can start being standardized. Through this process, a positive feedback loop is created in which promoting the use of data leads to higher resident satisfaction, which in turn leads to even greater use of data.

14.2.3 Key Factor 3: Citizen Participation

Active citizen participation is crucial to delivering smart-city programs oriented toward people's well-being. To encourage citizen participation in smart-city programs, spaces and opportunities for participation must be provided and these must be tailored to the needs of locals and the features of the community. Therefore, there is a need to devise an appropriate plan that takes these matters into account. Accordingly, to define the maturity levels for the citizen participation factor, we backcast four levels from a scenario in which citizens participate actively in smart-city programs (Fig. 14.3).

Level 0 represents a situation in which there is no opportunity for citizens to express their views about or engage in the co-creation of smart-city programs.

Level 1 represents a stage where there is public comment and other opportunities for the public to receive briefings on and express their opinions on smart-city goals and programs. It is also a stage where a locally rooted approach that will engender co-creation (a deeper level of participation) is investigated.

Level 2 represents a stage where sporadic workshops are held to identify specific themes for the smart city and create opportunities for citizens to engage in co-creation.

Level 3 represents a stage where there are opportunities for citizens to engage in co-creation in relation to themes pursued in the smart-city project and where such opportunities function as a basic mechanism in the smart-city. This stage corresponds to a living lab (see Chap. 7). "Functioning as a basic mechanism" means that people organizing smart-city projects are increasingly trying to create opportunities for co-creation and are using methods such as living labs to promote co-creation for a range of smart-city themes.

Level 4 represents a stage where the actions taken during level 3 are being applied and improved on an ongoing basis. At this level, difficulties may be encountered in bringing on board the organizations that should be engaging in the co-creation or in attracting members of the public to actively participate. In such cases, the design of your approach to co-creation should be amended or rethought as necessary. This is also the level where the actions should be made sustainable by building rapport with local members of the public and members of staff and by securing budgetary allocations.

To normalize a situation in which citizens are one of the parties engaging actively in co-creation for smart-city programs, more opportunities for local residents to engage in the smart-city project need to be created and the public's participation gradually deepened so that they will gradually gain increasing experience in being on the side that is creating the smart-city program. If level 1 involves the conventional approach to this end, namely, one-way briefings, then level 2 is where you gradually incorporate resident perspectives and experiment with ways to achieve higher quality participation and unleash creativity. To design a co-creation program that will be acceptable and accessible to the public, the goals and outcomes of co-creation need to be confirmed as well as what services are already under development. Participation opportunities to meet the needs of the target demographics can then be tailored to fit. For example, the frequency, time, number of meetings, and format of co-creation will vary depending on whether the target demographic is elderly people or people who are parenting children. It is no easy task to design co-creation spaces that will be acceptable and accessible to the target demographic, but with an accumulation of practices in level 2, more experience can be gained and the ability to work toward the goal of level 3, to design a flexible and inclusive co-creation program covering a range of themes in the smart-city project, can be improved. Hence, at level 2, we must engage in trial and error to build up experience in co-creation together with members of the public.

14.2.4   Key Factor 4: QoL-Based Assessment

The growing global need to measure and assess the impact of smart-city programs on people's QoL and well-being necessitates a clear quantitative representation of these outcomes. To achieve this, it is essential to define the metrics to be used and the data to be tracked. The maturity levels for this factor reflect the extent to which a system utilizing human-centric metrics has been developed to evaluate smart-city programs and the city's status (Fig. 14.3).

Level 0 represents a preliminary stage where no effective method exists to assess whether smart-city programs are enhancing the QoL, or if a method does exist, it is not recognized or utilized. At this level, the only assessment conducted is the evaluation of the program's progress toward a one-dimensional goal.

Level 1 represents a stage where research is conducted into an assessment method that focuses on residents' subjective experience. At this level the factors that strongly correlate with well-being are identified in the target community Smart City Institute Japan. These factors could be derived from the publicly available results of the community well-being scale developed by the Smart City Institute Japan and used in the Digital Agency's digital garden city nation strategy (Smart City Institute Japan 2023) or from Active QoL (see Chap. 8). These factors can then be used to clarify the strengths and issues of the community. This level also represents a state in which the relevant policies and smart-city programs have been selected to assess or new programs have been designed to be assessed in the future.

Level 2 represents a stage at which resident QoL or well-being surveys are utilized to assess smart-city programs or the status of the city at fixed time points. At this level, assessments begin by defining the target data and designing the survey to align with the objectives of the assessment and the subjects being evaluated. Additionally, data are collected and analyzed to evaluate the outcomes of the programs and to identify emerging issues. Based on these findings, a PDCA cycle is implemented, which includes designing future programs.

Level 3 represents a stage where the assessment cycle tested at level 2 is now integrated and functioning as a basic smart-city mechanism. This level also represents a stage of continuous data collection for assessing disparate programs and measuring changes across the short and long term.

Level 4 represents a stage where you are continually improving the level 3 processes. At this level, you periodically reform and refine the programs and mechanisms as part of the PDCA cycle to make them work more effectively.

The aim at this level is to use data to check whether the smart-city programs are contributing to people's QoL. To that end, the key element of level 3 is needed: a system of assessment tailored to the community. This system will involve analyzing the community-specific factors correlated closely with QoL to identify what community issues must be addressed urgently, what smart-city interventions might prove effective, and what programs should be assessed. Remember, the metrics selected must be impartial, incorporate resident perspectives, and must be open to the public; if metrics are used that are based only on the smart-city operator's perspective, there is a danger that the assessments will produce partial, biased results. Clarity of purpose is also important; at level 1, there must be clarity concerning why the smart city was assessed. Depending on the purpose, the reason is that the parameters of assessments will differ, including the metrics to be used, the geographic scope of the data sample and the period, frequency, and method of data collection. Generally speaking, there are three broad purposes for assessing a smart city:

1. To clarify the community's features and issues and present residents with evidence about the state of the community
2. To prepare basic data that can be used to examine ways to expand the community's features or address its issues
3. An impact assessment: To see how a proposed intervention would affect the community

In the case of (1) and (2), we would be looking at residents' QoL or well-being from a macro-perspective, in which QoL or well-being is defined as a feature of the community. On the matter of frequency of data collection, a one-off collection will probably suffice. If data are then collected every year or so, that would be even better. As for the size of the sample, the data collection could be made more efficient by adding the metrics of interest to a national government or local government survey conducted on a large population, as these surveys, with their large sample size, enable comparisons of subareas within the survey area. In the case of (3), because the purpose is to measure the impact of the intervention, we would be looking at residents' QoL or well-being from a micro-perspective, in which changes in behavior or satisfaction level are the focus. For a granular assessment that can track the dynamic progress of a smart-city program, think about eventually using an assessment such as Active QoL (see Chap. 8) in which residents' everyday activities and psychological states are tracked via behavior logs inputted on a smartphone or data from wearable sensors. Bear in mind, though, that wearable devices, by their nature, have disproportionate ownership and that devices other than smartphones are not yet widely available. For these reasons, limit the focus to a particular service or to a particular area.

14.2.5 Key Factor 5: Human Resource Development

Smart-city human resources do not grow by itself. It will require knowledge, skills, underlying ideas, and negotiations that differ markedly from existing business routines. An external agency can be hired to deal with human resource development, but the results will prove unsatisfactory unless the agency understands what new business processes are required. Thus, if the aim is to create a sustainable smart city, a conscious effort must be made to develop a workforce with the right knowledge and skills.

Making a conscious effort to develop a workforce means building a human-resource-development system. Accordingly, the maturity levels for this factor represent progress in the construction of such a system (Fig. 14.3).

Level 0 represents a stage where nothing has been done; no plans for human resource development are in place and staff are left to fend for themselves.

Level 1 represents a stage where research is underway. The criterion for this level is whether one is examining precedents and best practices for human-resource-development programs and whether there is consideration of how the best practices could be incorporated into your organization.

Level 2 represents a stage where human-resource-development programs are trialed. At this level, there is sporadic implementation of a system of human resource development with a mixture of external training programs and internal training.

Level 3 represents a stage where a human-resource-development program is standardized. The criterion for this level is whether the kind of workforce one wants to cultivate has been defined, and the best program for this vision selected, adopted, and standardized.

Level 4 represents a stage where the program and its standardized processes are continuously improved. At this level, the processes set up during level 3 is periodically reviewed and refined, and a budget and organizational framework are in place to sustain such continuous improvement.

A key focus in the above maturity levels is having a strategic approach to human resource development. The kind of workforce needed will differ depending on the organization, and the workforce cannot master overnight knowledge and skills that differ markedly from those used in prior business processes. A strategic readiness for a fluid workforce, in which transfers and resignations may be common, is also needed.

How can sporadic investment in human resource cumulate in an embedded human-resource-development system? Consider the following five pointers.

First, the kind of workforce to cultivate has been defined—set the objective. See Chap. 9 for insights about the kind of skillsets and workforce size that may be required.

Second, the "who" and the "where" of the training need to be defined. In anticipation of workforce fluidity in the future, there must be an idea of who in the organization needs to be trained, and also where (into which positions) human resources should be allocated.

Third, there needs to be a training methodology. For example, have employees attend training programs to expand their knowledge, have them engage in on-the-job-training (by working in a highly skilled team) to refine their practical skills, or organize internal workshops to expand the range of people with skills and experience related to smart cities. For the architect role, which requires a particularly advanced level of expertise, consider headhunting and bringing in human resources from outside, as it might take too long to train up the human resources from within the ranks.

Fourth, provide a framework to organizationally embed the program. If the program involves employees attending a training course, it may be a good idea to subsidize the course fees, in which case it will be necessary to think about how many people will be attending each year. One should also organize internal workshops where the course attendees can share the outcomes of their learning with colleagues; this will benefit the attendees themselves as well as their colleagues.

Fifth, keep up to date with the latest developments. Technological and social trends related to smart cities move fast. An environment needs to exist that is always accessing the latest news about programs in other smart cities, thereby encouraging a constant updating of intelligence. To that end, join external information-sharing networks and have processes in place for keeping intelligence up to date.

Hopefully, these pointers will help designers reach level 3 as soon as possible. Human resource development has synergistic effects; the more the workforce is trained, the more colleagues interact, share knowledge, and learn from one another.

14.2.6   Factor 6: Data Ecosystem

The aim in cultivating a data ecosystem is to increase the numbers of data providers and data users exchanging data between organizations in the community and sustain a situation in which a plethora of data-driven services deliver value to residents and organizations. For this, it will be necessary to introduce a pricing mechanism (not just monetary pricing); there is also a need to keep introducing new services and reforming existing ones (i.e., achieve service metabolism) to dynamically adapt to needs that change over time.

Accordingly, the maturity levels for this factor represent the degree to which a body of precedents and best practices related to interorganizational data use, the degree to which such data is sustainably entrenched (by introducing a pricing mechanism, for example), and the extent to which the services driven by the data are renewed or replaced according to demand is necessary (Fig. 14.3).

Level 0 represents a stage where there is no interorganizational data sharing in smart-city programs.

Level 1 represents a stage where you are investigating needs, and seed capital that could match such needs, in relation to interorganizational data use.

Level 2 represents a stage where interorganizational data use is being trialed. At this level, you have trialed interorganizational data use in relation to a particular smart-city theme and verified these use cases.

Level 3 represents a stage where multiple services have started delivering the expected benefits and revenue such that the services are now sustainably entrenched. It also represents a stage where you have derived from high-quality use cases effective strategies for supporting data use and where these strategies are now integrated and functioning as a basic mechanism in smart-city management.

Level 4 represents a stage where mechanisms are working to adapt services to changing needs. At this level, services driven by shared data are scrapped and replaced as necessary in accordance with feedback from residents and other stakeholders, and mechanisms are operating to sustain the level 3 activities in a form that adapts to changing needs.

A criterion we used when defining level 4 was that the data ecosystem must have prospects for long-term growth rather than just amounting to one-off or superficial data exchanges. While it is important to lay down data infrastructure and have a robust system of testing and verification, what really matters is that the services are established as viable businesses and that they are developed in an integrated manner, as part of the overarching management of the data ecosystem.

We offer the following two pointers for cultivating an ecosystem that can make the people-centered city sustainable. First, in developing the ecosystem, an approach should be taken that focuses on use cases. Laying down data infrastructure will go some way in facilitating data use, but it is insufficient by itself to proliferate use cases. Once the needs and the seed capital that could match such needs in relation to interorganizational data use have been identified, on the focus can be on creating simple, small-scale use cases (those that can be easy to implement from scratch). These simple cases will pique the interest of stakeholders, leading to a buildup of use cases in a snowball effect. In other countries, we can find cases in which open-data processes such as ideathons and hackathons have culminated in apps used by residents.

The second pointer is to establish a pricing system for data-driven services, which often present challenges in immediate value quantification. In many instances, it is not feasible to determine the monetary value upfront. A comprehensive perspective on pricing is essential when designing sustainable services. For example, if a service contributes broadly to community welfare—such as enhancing street safety—it may be beneficial to use infographics that illustrate these benefits. This visual representation helps the public understand the costs associated with these benefits more clearly. In Chap. 10, our research findings indicated that many organizations were participating in data-driven services primarily to contribute to the community rather than for direct financial gains. Therefore, it would be prudent to communicate to the public which organizations are involved in these services and clarify their community-focused intentions.

Hopefully, these two pointers will help in reaching level 3 of the data system factor as soon as possible.

14.3   Stepwise Implementation Based on the Key Factor Levels

We believe that the four maturity levels delineated for each of the six key factors provide a systematic framework for developing a people-centric smart city within the target community. This framework facilitates the identification of current progress and determination of subsequent objectives.

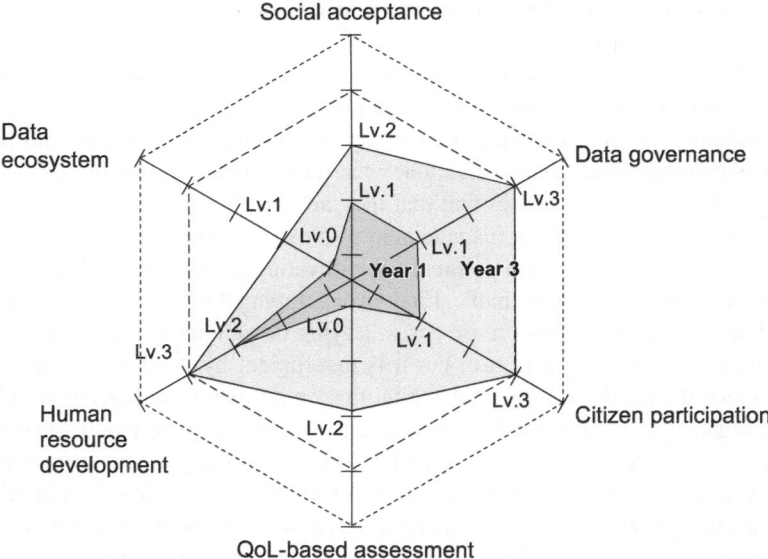

Fig. 14.4 Example of using a radar chart to plot your progress in the maturity levels for the six key factors as part of a self-analysis

Questions such as "Which factors have shown substantial progress? Which have seen little advancement? Which have not been addressed at all?" can be clarified by utilizing a radar chart. This chart plots the maturity levels of each factor, exemplified in Fig. 14.4. These levels serve as a common frame of reference, simplifying collaboration with team members and other stakeholders. They assist in conducting self-analyses to gauge the extent to which initiatives contribute toward a people-centric smart city and in setting long-term goals. Although the challenges of achieving these key factors may seem daunting, we advocate for a stepwise approach to progression: starting initiatives at level 1, experimenting and refining them at level 2, and subsequently, clarifying the path forward.

Smart cities are not built overnight; they are the result of long-term initiatives. These developments involve multiple stakeholders and a wide range of technological innovations. While it is necessary to establish maturity levels and draft an implementation roadmap for each technology, the focus must consistently remain on the primary goal: to create a people-centric smart city that enhances well-being. Progressing through these levels often involves a steep uphill climb, which underscores the importance of having a common frame of reference. This framework should be similar to the SAE levels of automation, defining distinct levels of maturity that facilitate discussion and collaboration at each stage, and aid in identifying challenges that no single community could surmount on its own. To all stakeholders engaged in developing a human-centric smart city, we recommend adopting our system of maturity levels for the key factors. We urge you to utilize this system to methodically advance the implementation of these key factors.

14.4   Key Factors as Infrastructure for a Digital Society

This book has proposed, and elaborated on, the necessity and importance of six key factors essential for developing a people-centric sustainable smart city. Central to these factors is the understanding that smart cities cannot be constructed overnight. Instead, all relevant public and private sector organizations must continuously manage and advance the process of smart city creation, utilizing both the Society 5.0 reference architecture (Cabinet Office 2020) and the smart-city reference architecture (Cabinet Office 2023).

The term "architecture," adopted into Japanese as the loanword *ākitekucha*, traditionally refers to the physical construction of buildings or to the structures themselves. However, as the introduction to this book clarifies, smart cities are distinct from physical buildings in a crucial aspect—they are never truly completed.

Typically, a building begins to be utilized only after its construction is completed; for example, houses are inhabited only after they are fully built. Subsequently, these buildings are maintained (through maintenance work or usability improvements), suggesting that their completion in terms of management is a continual process. This concept is even more applicable to complex urban environments, which are designed for perpetual evolution and are constantly undergoing construction and redevelopment. Similarly, smart cities are dynamic entities, continuously evolving and adapting to new technologies and resident needs.

Recall what it entails for a smart city to be people-centric: It involves continuously creating and effectively managing services in response to the evolving needs of its inhabitants (users, members of the public). While this may appear to be an unending, Sisyphean task, it is a necessary path of creation and renewal that must be followed to enhance the livability of the city or community.

To create and renew livable cities and communities suited for the emerging digital society, we need clear signposts—common infrastructure and benchmarks—to guide our progress. In the twentieth century, cities were supported by two types of social infrastructure. The first type was network infrastructure, which included rail, road, electricity, gas, and water networks. The second type was software-based digital social infrastructure, which encompassed amenities with significant public dimensions (such as schools, hospitals, parks) and the organizational frameworks necessary for their use (education systems, healthcare systems, etc.). Although it was uncommon for the maturity levels of these infrastructures to be quantified in clear numerical formats, general standards existed to provide a reasonable assessment of the city's or community's status in terms of education, healthcare, and similar services. Similarly, today's smart cities require a new form of software-based digital social infrastructure that enables local communities—the central actors in smart cities—to efficiently navigate the path of creation and renewal. This necessity inspired the proposal of the six key factors.

Twentieth-century societies were supported by social systems and applications, but cities and communities within the digital society, being at a more advanced stage, will rely on a new form of social infrastructure. This infrastructure will consist of innovative hardware networks such as ICT and high-speed communication networks, alongside their social systems and applications. However, it is critical to avoid focusing solely on efficiency and economic benefits to the extent of becoming overly dependent on hard technology; progress should be people-centric and sustainable. This principle has been consistently emphasized throughout this book. To ensure that smart cities develop in this manner, the six key factors must be integrated as essential components of the infrastructure in a digital society, thereby enhancing the well-being of the community. This is the book's concluding assertion and recommendation.

The six key factors should be communicated within our cities or communities along with a set of maturity levels (ranging from 0 to 4) for each factor. This framework will allow residents to discern the current developmental stage of their community and understand the subsequent phases, thus setting a clear path toward developing a people-centric, sustainable smart city. Moreover, the smart-city initiative is a journey of creation and renewal, one that necessitates active participation from both citizens and residents. For these reasons, we encourage the utilization of the six key factors and their associated maturity levels.

References

Cabinet Office (2020) Smart City Reference Architecture White Paper, 1st edn. In: Cross-ministerial Strategic Innovation Promotion Program (SIP) Second Phase, Big-data and AI-enabled Cyberspace Technologies/Smart City Architecture Development/Smart City Architecture Design and Promotion of Related Verification Research (Released on March 31, 2020). https://www8.cao.go.jp/cstp/stmain/20200318siparchitecture.html. Accessed on March 7, 2024.

Cabinet Office (2023) Sumāto Shiti: Rifarensu ākitekucha howaito pēpā [Smart city: Reference architecture white paper], 2nd edn. In: Cross-ministerial Strategic Innovation Promotion Program (SIP) Second Phase, Big-data and AI-enabled Cyberspace Technologies/Smart City Architecture Development/Smart City Architecture Design and Promotion of Related Verification Research (Released on August 10, 2023). https://www8.cao.go.jp/cstp/stmain/20230810smartcity.html. Accessed March 7, 2024.

International Organization for Standardization (2017) ISO 37153:2017 Smart community infrastructures– Maturity model for assessment and improvement https://www.iso.org/standard/69225.html. Accessed April 30, 2024.

International Telecommunication Union (ITU) (2020) Y.4904: Smart sustainable cities maturity model, https://www.itu.int/rec/T-REC-Y.4904/en. Accessed April 30, 2024

SAE International (2021) Taxonomy and definitions for terms related to driving automation systems for on-road motor vehicles, J3016_202104, 30 April 30 2021 https://www.sae.org/standards/content/j3016_202104/. Accessed April 30, 2024.

Smart City Institute Japan (2023) Chiiki kōfukudo (well-being) shihyō [Community well-being indices]. https://www.sci-japan.or.jp/LWCI/index.html. Accessed April 30, 2024

Strategic Conference for the Advancement of Public and Private Sector Data Utilization, Strategic Headquarters for the Advanced Information and Telecommunications Network Society (2021) Kanmin ITS kōsō rōdomappu: Kore made no torikumi to kongo no ITS kōsō no kihonteki kangaekata [Public-Private ITS Initiative/Roadmaps: Past initiatives and basic approach to future ITS scheme] June 15, 2021 https://cio.go.jp/sites/default/files/uploads/documents/its_roadmap_20210615.pdf. Accessed April 30, 2024.

Chapter 15
Dialogue: Creating the Future Through Dialogue: How to Achieve Society 5.0 and Deliver Well-Being Without Exceeding Planetary Boundaries

Teruo Fujii and Toshiaki Higashihara

Abstract Smart city initiatives are underway in various locations in Japan, serving as preliminary explorations of Society 5.0, a people-centric super-smart society envisioned in the government's fifth Science and Technology Basic Plan of 2016. Alongside this, the University of Tokyo and Hitachi have been engaging in joint research on two topics in particular that are aimed at achieving Society 5.0: the creation of sustainable and people-centric smart cities and the formulation of a vision for the energy systems of the future. This scheme for collaboration between industry and academia being undertaken at the Hitachi-UTokyo Laboratory has been recognized as a model for the harmonious collaborative creation (co-creation) that is essential to overcoming societal challenge. Globally, the situation is becoming increasingly complex, with worsening climate change and a changing international order. Japan, meanwhile, is facing the new reality of depopulation. Combining people's well-being with a society that can stay within planetary boundaries will be crucial to overcoming these challenges. In this chapter, Teruo Fujii, President of the University of Tokyo, and Toshiaki Higashihara, Director and Executive Chairman of Hitachi, Ltd., engage in a discussion that sheds light on how the University of Tokyo and Hitachi are addressing this task.

Keywords Society 5.0 · Well-being · Planetary boundaries · DEI (Diversity, Equity, and Inclusion) · GX

The above is an extract from "Hitachi Technology 2023: Technology & Innovation Foresights 2023," Hitachi Review Vol. 72, No.2 (June 2023).

T. Fujii
The University of Tokyo, Tokyo, Japan

T. Higashihara (✉)
Hitachi, Ltd., Tokyo, Japan
e-mail: office@ht-lab.ducr.u-tokyo.ac.jp

© The Author(s) 2025
Hitachi-UTokyo Laboratory (H-UTokyo Lab.), *The Architecture of "Society 5.0"*,
https://doi.org/10.1007/978-981-96-2929-9_15

Teruo Fujii, President of the University of Tokyo (left). Toshiaki Higashihara, Director, Executive Chairman, Hitachi, Ltd. (right)

15.1 Changing Environment for Both University and Company

Higashihara I am delighted to have this opportunity to discuss Society 5.0 with you today, as it is one of the research topics being addressed by the Hitachi-UTokyo Laboratory.

Following your appointment as President of the University of Tokyo in April 2021, you published new guiding principles of the university called UTokyo Compass that came out in September of that year. Can you please tell us what you were seeking to achieve by this?

Fujii UTokyo Compass was titled "Into a Sea of Diversity: Creating the Future through Dialogue" and highlighted three core values: "creation through dialogue," "diversity and inclusion," and "a university for everyone in the world." Along with a global agenda that includes climate change and the pandemic, humanity is also faced with the threat to multilateralism posed by Russia's invasion of Ukraine. In such a world, universities have an even greater role to play as places where people from many different walks of life can engage in dialogue.

Dialogue is the act of trying to learn something unknown. To discover and understand something unknown, we need to pose questions. By sharing and thinking about common questions through dialogue, we can build mutual understanding and trustful relationships among people. Universities can serve as a place for creating knowledge and building the future, discovering solutions to a variety of different challenges through dialogue, not only within academia, but also with people from outside or overseas who come from different backgrounds. When we fulfill this mission, I believe that universities are contributing to international society. I am conscious of how co-creation between industry and academia, as exemplified by the Hitachi-UTokyo Laboratory, provides a framework for creating the knowledge to take on difficult challenges through dialogue.

Higashihara Dialogue is also a key concept in today's corporate activities. This year is the 113th since Hitachi was founded, and over the 100 years or so of our business, we have largely pursued a "product out" approach in which factories have played a central role. In recent years, however, that approach has undergone considerable change as co-creation with customers has come to play an ever more important part. Co-creation means sharing a vision and goals with the customer, identifying the challenges that need to be overcome to achieve those goals, and making use of digital technologies to find solutions. To borrow your own words, it is about dialogue as a business model.

While products remain important, these increasingly large and complex challenges cannot be resolved by the conventional approach to manufacturing. Along with the immediate challenges facing them and their customers, the companies of today also need to take a broader and more long-term perspective that incorporates the resolution of societal challenges into their businesses. Doing so calls not only for co-creation with individual customers, but also for working through consortiums of companies; collaboration across industry, academia, and government; and dialogue with non-profit organizations (NPOs) and communities.

In the "product-out" era, value was delivered through productivity-boosting efficiency gains and profit-enhancing cost savings. Nowadays, in contrast, what customers are seeking is changing toward forms of value that encompass societal challenges. For example, rather than just price and quality, when choosing products and services, customers are starting to prioritize things like the use of recycled materials and whether materials procurement and production respect human rights and are not damaging to the environment. With planetary boundaries (global-scale environmental issues) and well-being (individual happiness based on leading a life that is both mentally and physically healthy) becoming key considerations for business, I have come to recognize the importance of how co-creation between industry and academia brings a transformation in our thinking and an appreciation of new ways of looking at things that incorporate these ideas.

15.2 Acceleration of Green Transformation

Fujii While I believe that climate change represents the greatest societal challenge that we currently face, what we have now is better described as a "climate crisis" whereby continuing along our current path will result in action being too little, too late. In recognition of the need for a fundamental reorganization of social and economic systems as well as the structure of industry, we highlighted green transformation (GX) as one of the key pillars of the UTokyo Compass action plan, with a three-tiered approach that addresses the issue at the global, national, and university levels.

At the global level, the Center for Global Commons at the University of Tokyo[1] has played a central role in developing and publishing the "Global Commons Stewardship Index".[2] The objective is to stimulate international policy debate on how to safeguard the global commons and to encourage behavioral change at various levels of society in individual nations.

At the national level in Japan, we established the Energy Transition Initiative—Center for Global Commons (ETI-CGC) as a platform for investigating how Japan can make the energy transition to carbon neutrality by 2050, with participation by senior managers from 13 Japanese companies, including yourself. The intent is to develop Japan's pathways for achieving both carbon neutrality and well-being, including the formulation of policy proposals.

At the university level, we are playing our part in pursuing the goal of net-zero by 2050, participating as an academic partner in "Race to Zero," an international campaign launched by the United Nations Framework Convention on Climate Change (UNFCCC). We formulated and published our action plan for achieving this in October 2022, entitled "UTokyo Climate Action." Based on this plan, students, faculty, and staff members will work together to strengthen sustainability at the university.

Higashihara Government action on GX has included the July 2022 inaugural meeting of the GX Implementation Council chaired by the prime minister. Against a backdrop of ongoing energy supply concerns, this included work on what needs to be done to achieve carbon neutrality by 2050 and the creation of a roadmap for the JPY150 trillion of combined public and private investment planned over the next 10 years.[3]

As for industry, the Japan Business Federation (Keidanren) is seeking to achieve "Society 5.0 with Carbon Neutral," its term for a new economy and society that features fundamental change in how energy is sourced, innovation in the processes of production, wider adoption of innovative products for decarbonization of the transportation and consumer sectors, public behavior change, and a transformation in living practices.

---

[1] An organization launched in 2020 with the goal of serving as a facilitator, guiding the reform of societal and economic systems through co-creation with leaders from a wide range of fields in order to secure the earth as a global commons, humanity's common property. With the aim of building a sustainable future for both people and the planet, the center is working on a common international intellectual framework for managing the global commons (global commons stewardship). Based on this framework, the center is also working with a diverse range of stakeholders to encourage the transformation of societal and economic systems.

[2] An overall indicator that is used to score nations for their impact on climate change and biodiversity and to encourage action by serving as a basis for comparison. Jointly developed by the Sustainable Development Solutions Network (SDSN) of the United Nations, Yale University, and the Center for Global Commons at the University of Tokyo, indicators have been published for more than 100 nations to date.

[3] A draft of the "Basic Policy for the Realization of GX" with 10-year roadmap was presented at the Fifth GX Implementation Council held on December 22, 2022.

15.3 Taking an All-Encompassing View of Environmental Problems

Higashihara When considering fundamental changes in the sources of energy, it is vital to differentiate between the short- and medium-to-long-term future. While an expansion of renewable energy is essential in the long term, Japan will face electricity shortages in 2030 unless it makes progress on restarting its nuclear power plants. To ensure security of supply, we need to structure our electric power supply system in a way that treats nuclear power generation as a significant source of baseload power. In terms of thermal power generation, we need to establish a clear pathway toward reducing it, while also reducing carbon dioxide (CO_2) emissions by making greater use of alternatives such as hydrogen or ammonia, which can be mixed in with existing fuels. At the same time, there is also a need for longer-term action in the form of fundamental research into carbon capture and storage (CCS) and carbon capture utilization and storage (CCUS) so that it will be ready for practical implementation in the decade starting around 2040.

On the demand side, wider adoption of distributed power sources with the use of batteries or artificial intelligence (AI) for energy management will likely play a key role, facilitating the use of renewable sources of energy by communities, factories, offices, and families together with energy efficiency measures for saving electricity.

Carbon neutrality is far from the only challenge, however, and it is also important that we think about how to sustain global systems and the ecosystem. We live in a time when we need to be thinking about the impacts we have on the environment in comprehensive and global terms, encompassing every stage from materials and other resources to the factory and office, transportation, use of products and services, and their ultimate disposal and recycling. We need to be deploying technologies that can reduce the load on the environment, not only in Japan, but also elsewhere such as in Asia and the Global South.

Fujii As you say, it is vital to adopt an all-encompassing view and address the issues in a scientific manner. The information disclosure framework for organizations involved in sustainability has been expanded from the Taskforce on Climate-related Financial Disclosure (TCFD) to the Taskforce on Nature-related Financial Disclosures (TNFD), calling for reporting, not just on CO_2 emissions in the supply chain, but also on assessments of the risks and opportunities posed to the natural environment and biodiversity across all steps in the value chain. Considering issues such as those associated with food or human rights in procurement and production, there is a need for further research into how we can monitor the activities of companies and other organizations in all their different facets together with the collection of this data and the calculation of indicators.

Higashihara Transparency and analysis are essential if effective action is to be taken. This in turn calls for data collection and analysis platforms. With the European Union (EU) already working on its Gaia-X project to establish integrated infrastructure for the coordination of data across different companies, I feel that Japan, too, should be talking about open platforms for the integrated collection and coordination of data between companies in Japan and Asia that will facilitate the provision of services.

15.4 Need to Adopt a User Perspective

Higashihara Alongside planetary boundaries, Hitachi also sees "well-being" as an important consideration, one that is highlighted as a key concept in the people-centric super-smart society of Society 5.0. Smart cities are seen as providing a preview of Society 5.0. At Smart City Institute Japan,[4] meanwhile, in which Hitachi is participating alongside academics from the University of Tokyo and where well-being is defined as living a fulfilling life, support is being given to work on smart cities that seeks to improve on this measure. The institute has also developed the Liveable & Well-Being City Indicators for use in activities taking place around the country. Intended specifically for Japan, these indicators can be applied to urban developments that enhance the well-being of community residents.

While people tend to think of smart cities as greenfield developments that start from a blank slate, brownfield projects in which digital technology and data are put to work to transform existing cities are likely a more realistic solution. One notable example comes from Kakogawa City in Hyogo Prefecture where an investigation into the potential for installing surveillance cameras to reduce the crime rate was prompted by community feedback to also incorporate a monitoring service for children and the elderly that works by using beacon tags,[5] thereby improving the town's livability. The requirements for establishing a smart city are: (1) leadership, (2) clearly defined objectives and key performance indicators (KPIs) for the planned activities, and (3) community involvement. In the case of Kakogawa City, however, three key elements that came together to make their project a success were, I believe: (1) the mayor, (2) the installation of a crime prevention and monitoring service, and (3) the adoption of a feedback scheme that can collect a wide range of community views by means of a website.

An advantage of brownfield projects is that they are able to make use of community feedback while changes are being put in place. In recent years, a growing

---

[4] A not-for-profit organization established to promote the expansion and enhancement of smart cities in Japan.

[5] A public–private project being jointly undertaken by Kakogawa City and a number of private-sector businesses. The project has installed surveillance cameras equipped with beacon tag detectors in the area around an elementary school and along the paths taken by children going to and from school. The system is used to monitor the children and others, including elderly people suffering from dementia. It works by updating a smartphone app belonging to the person's caregiver or family with details of their movements whenever they pass near one of these detectors while wearing the tag.

number of people have come to treat societal challenges as matters of personal concern to the extent that they are prepared to involve themselves in urban development. I am hopeful that brownfield smart city projects will prove to be the catalysts that initiate the transition to Society 5.0. In the private sector, likewise, we want to play our part in this movement through means such as digital technology.

Fujii As a university, being a place where different types of people can come together is one of our defining characteristics and, as such, I hope that we can serve as a place for community participation and for connecting stakeholders together.

For Society 5.0, it is important to clarify who we are talking about when we refer to it as "people-centric." As you mentioned earlier, whereas the evolution of twentieth-century industry was more about the things that mattered to producers, the progress of digitalization since we entered the twenty-first century has placed greater weight on the perspectives and values of the people who use products and services.

This change now informs the debate about smart cities, where a switch in focus toward user considerations is called for in fields like social infrastructure and in public services such as healthcare and education. Those in healthcare are asking themselves what constitutes value for patients, while for us in the education sector, we need to be rethinking what is needed if students are to study the things they really want to study. This is a time for the providers and consumers of services to be getting together to engage in genuine dialogue and to think about issues of concern and what constitutes value in those services.

15.5 Treating Issues as Matters of Personal Concern

Fujii For myself, I have spent a lot of time studying the ocean and have been engaged in the Ocean Monitoring Network Initiative (OMNI). I have launched this project to perform large-scale oceanographic surveys using low-cost sensor systems. To achieve this, the project has drawn on design capabilities to develop sensors and data platforms for collecting a variety of oceanographic data such as water temperature and salinity, with involvement by the general public as well as researchers. Participation in the project fosters a sense of personal connection with the ocean environment, with activities including workshops for elementary, junior high, and high school students and helping them come up with ideas for sensing devices. Having more people collecting observations increases the volume and resolution of the oceanographic data while also providing an opportunity for personal behavioral change as people gain a shared awareness of the problems. I hope that this project will serve as a model for using dialogue with the community as a means of overcoming challenges.

Rather than being conferred by other people, I believe that well-being is something we acquire when we voluntarily engage with each other to make society better.

Higashihara That is right. Because well-being means different things to different people, if everyone pursues happiness in their own way, they may end up just getting in each other's way, like trying to inflate a lot of balloons in the same small space. What is called for, rather, is empathy and to consider the viewpoints of others. The same applies to planetary boundaries, a problem where it is important that we think in terms of win–win rather than zero-sum outcomes.

The best overall outcomes can be achieved by local government, companies, universities, and residents adopting this approach to working with one another to address local issues, with these communities working together seamlessly based on Japan's Vision for a Digital Garden City Nation. It is when this happens that I believe well-being will be improved.

Fujii Working with the local community is also important for us. In November 2022, we entered into a comprehensive partnership agreement with Wakayama Prefecture. The aim of the agreement is to help resolve local issues and create a distinctive regional society through academic research and the exchange and training of personnel.

While this is the third such agreement, we have reached at the local government level, the earlier ones being with Mie and Fukushima prefectures, a number of our faculties were already doing work in Wakayama. The Institute of Industrial Science to which I belonged has set up a laboratory at Kada in Wakayama City to conduct studies aimed at regional revitalization, while the Research Center for Advanced Science and Technology partnered with Kongobu-ji, the Main Temple of the Koyasan Shingon Sect, as well as with Koyasan University and Koya Town to hold the Koyasan Conference. Similarly, our Graduate School of Humanities and Sociology has entered into an agreement with Shingu City involving the establishment of a satellite facility and the hosting of the UT Jimbun-Kumano Forum. The graduate school has also entered into an agreement with Kitami City in Hokkaido to engage in joint work in which Shingu City will also participate, with plans to go deeper into areas such as multiregional cross-cultural studies.

We have also embarked on a "field study-based partnership project between local governments and UTokyo." This involves students spending time at local government agencies to conduct on-site studies of local issues and come up with ways of resolving them with assistance from faculty and staff members. During the 2022 academic year, we worked with 19 such local governments and communities at locations across Japan.

15.6 Improving Diversity, Equity, and Inclusion Through Mutual Understanding

Higashihara Diversity and inclusion are both key considerations when seeking to create a people-centric society, and together with equity, these concepts have been a focus of much attention over recent times, collectively known by the abbreviation

"DEI." For Japan to retain its vitality as it confronts the reality of depopulation, it is essential that both the private and public sectors promote globalization and diversity more than ever before. To this end, Hitachi has declared a goal of increasing the percentage of female and non-Japanese executive and corporate officers to 30% by 2030.

Equity, being about the rectification of imbalances with respect for difference, is not the same as equality. Rather than something that can be expressed in simple rules, it is about formulating rules that are consistent with an understanding of people's individuality and of local culture and history, or of providing support and creating a level playing field that leads to a society that leaves no one behind. In other words, it is about mutual understanding. It means first understanding and accepting the values of other people, and then having them understand your values. While it may seem trivial, I believe that order is crucially important.

Fujii I agree. Building relationships of trust through dialogue demands both understanding and respect for others as you mentioned. As I spoke about earlier, the University of Tokyo is seeking to become "a university where anyone in the world would want to come and join." Being accepting of people from a range of different backgrounds is important in academia just as it is in business. In order to raise the level of our research, debate needs to encompass diverse viewpoints. DEI is implicit in the topics like Society 5.0 and well-being that we are addressing in our discussion and I believe that they constitute genuine value.

One of the ways in which universities need to respond to depopulation is by fostering and utilizing a diverse range of people so as to maintain national vitality. As part of such efforts, we aim to recruit 300 female teaching staff by the 2027 academic year. By raising the percentage of female faculty and supporting their activities, our hope is that this will also increase the number of female students. Likewise, increasing the number of international students should foster deeper understanding and allow them to develop closer ties to Japan while also helping to improve diversity domestically.

Disability inclusion is also essential. We have students who get around the university in wheelchairs and the Research Center for Advanced Science and Technology is engaged in an assessment and rehabilitation project whereby people with disabilities or illness study their own concerns, such as the difficulties they experience and the ways in which their illness manifests in their daily lives. Providing an inclusive environment for education and research not only promotes innovation, but also serves as a model for other organizations.

15.7 Responsibilities of Science and Technology

Higashihara Along with changes in attitude, the way in which DEI is achieved at a technical level also raises issues. For example, as long as they are able to operate the relevant devices, new digital realms such as the metaverse allow people to communicate, work, and study wherever they live and without the constraints of time, regardless of their age, gender, or physical condition. If not only sight and sound, but also other senses such as touch, smell, and taste can be provided as virtual experiences, then the possibilities could extend to the immersive study of history or the interactive acquisition of skills and expertise.

On the other hand, these new technologies have both advantages and disadvantages and come with ethical issues. Given that virtual reality is artificial, how far can we allow it to go? While Hitachi provides training to our researchers and engineers about the ethics of emerging technologies like gene recombination, genome editing, and regenerative medicine, there is also an urgent need to establish effective rules through a process that includes public debate and assessment.

Fujii The ethical, legal, and social issues (ELSI) raised by science and technology have themselves become a topic of academic study. Moreover, the EU has, in recent years, been addressing the concept of responsible research and innovation (RRI).

As exemplified by the issue of genetically engineered foods, new technologies can foster uncertainty and distrust in science when they first enter the public realm. To avoid this, rather than a closed debate among the ranks of scientists, it is important to engage in dialogue with the public from the early stages of research and development. While responsibility as a concept also embodies ethics, I see it as having two aspects when it comes to research, namely, the responsibility to "create" scientific knowledge and the responsibility to "utilize" scientific knowledge. Scientists and other researchers are called upon to fulfill both of those responsibilities, which is why we include "Promote Responsible Research" as one of the 20 goals in the UTokyo Compass.

15.8   Mutual Engagement for Development of Human Resources and Resolution of Challenges

Fujii Given a global agenda including the ethical questions of new technology as well as climate change and the new international order characterized by the Covid-19 pandemic and the invasion of Ukraine, a mountain of issues calls for our action. Today's discussion has reinforced for me how our ability to overcome these challenges is enhanced by companies like Hitachi and universities like ours working together to take them on.

As a university, providing students with on-the-ground experience is a particular emphasis of ours when it comes to addressing ever more complex challenges. We promote internships both in Japan and overseas in the hope that knowledge of the practical world will encourage students to make a personal commitment to action on these challenges. I also see an important place for startups when it comes to initiatives for addressing challenges. As many students want to be involved with startups in the social as well as the technology sphere, such as businesses that can make a social contribution on the ground in the Global South, for example, the university is looking for ways in which we can support them.

Higashihara While we spoke earlier about the university's role in developing talent, companies also have an important role to play in fostering people who can take on a global agenda. This is about having a personal commitment to addressing societal challenges from a global perspective and resolving those challenges in a way that brings people onboard based on mutual understanding. What I would like to see is for us to build a relationship in which, through mutual engagement, we can resolve societal challenges together. More than just internships, this could involve things like our people going to the university to learn or having university researchers come to us.

Fujii I very much hope we can do that. When it comes to the fostering of talent at universities, I am conscious of the need for us to build up our capabilities for acquiring understanding of other subjects as well as our own, or for engaging in team-based research with specialists from other fields. It is by acquiring such capabilities that our engagement with companies will deliver benefits.

Higashihara With a philosophy like that, the Hitachi-UTokyo Laboratory will have a more important role to play than ever. The pursuit of interdisciplinary research is an area worthy of particular attention along with incorporating public feedback into efforts to create Society 5.0, and we need to make further progress toward it serving as a forum for dialogue that encourages active debate and the exchange of views between companies and universities along with numerous other stakeholders. I look forward to its coming up with many good ideas that will help Hitachi toward our goal of using the resolution of societal challenges as a means of creating a society that delivers well-being to everyone without overstepping planetary boundaries. Thank you for your time today.

Teruo Fujii   President, The University of Tokyo
Received his Ph.D. in engineering from the University of Tokyo in 1993. Following work at the University of Tokyo Institute of Industrial Science (IIS) and the Institute of Physical and Chemical Research (RIKEN), he served as Director General of the IIS, Executive Director and Vice President of the University of Tokyo, and Executive Vice President (in charge of finance and external relations) of the University of Tokyo. He was appointed the 31st President of the University of Tokyo in April 2021. In September 2021, he published "UTokyo Compass—Into a Sea of Diversity: Creating the Future through Dialogue," a statement of the guiding principles of the university. Since March 2021, he has served as an Executive Member (part-time) of the Council for Science, Technology, and Innovation. His specialties are applied microfluidics systems and underwater technology.

Toshiaki Higashihara Director, Executive Chairman, Hitachi, Ltd.

Joined Hitachi, Ltd. in 1977 after graduating with a degree in electrical engineering from the Faculty of Engineering at Tokushima University. He obtained a Master of Science in Computer Science at Boston University in 1990. His past roles have included COO of the Information & Telecommunication Systems Group, President of Hitachi Power Europe GmbH, President and Representative Director of Hitachi Plant Technologies, Ltd., Vice President and Executive Officer of Hitachi, Ltd., President & CEO of Hitachi, and Executive Chairman & CEO of Hitachi. He took up his current position in April 2022.

Epilogue: Toward the Exploration of the New Frontier of Cyberspace

Atsushi Deguchi and Hideyuki Matsuoka

The Story So Far

In 2018, H-UTokyo Lab, an industry-academia collaboration organization, published the book of *Society 5.0*. This lab was established in June 2016 following an agreement between Hitachi and the University of Tokyo. It focuses on two core projects: the Habitat Innovation Project, which oversaw the publication of this book, and the Energy Project. The findings from these projects have been disseminated through our website, white papers, and other print media. Our academic-industrial research intensified in the fiscal year ending March 2018. During the first phase of our research (April 2017–March 2020), we shared our findings through various forums and in our publication *Society 5.0*. The second phase (April 2020–March 2023) saw a shift to predominantly online research due to pandemic restrictions. In October 2021, the Habitat Innovation Project hosted a well-attended forum at the University of Tokyo's Ito International Research Center (Ito Memorial Hall). The joint research discussed at this event was further refined and the conclusions were incorporated into this book.

The concept of Society 5.0 was initially outlined in the Government of Japan's fifth Science and Technology Basic Plan, which was approved by the Cabinet in January 2016 and subsequently presented to an international audience. The publication of Society 5.0 (H-UTokyo Lab 2020) has seen over 10,000 copies printed, indicating its significant role in promoting this vision. In March 2021, the Cabinet approved sixth STI Innovation Plan, elaborating on the Society 5.0 concept and

A. Deguchi
Department of Socio-Cultural Environmental Studies, Graduate School of Frontier Sciences, The University of Tokyo, Tokyo, Japan
e-mail: deguchi@edu.k.u-tokyo.ac.jp

H. Matsuoka
The Basic Research Center, Research & Development Group, Hitachi, Ltd, Tokyo, Japan
e-mail: hideyuki.matsuoka.ws@hitachi.com

© Hitachi-UTokyo Laboratory (H-UTokyo Lab.) 2025   243
Hitachi-UTokyo Laboratory (H-UTokyo Lab.), *The Architecture of "Society 5.0"*,
https://doi.org/10.1007/978-981-96-2929-9

implanting the idea that Society 5.0 serves as a future vision in which Japanese science and technology lead the way. Many initiatives have been launched that are aligned with the vision and that are based on nationally subsidized programs.

In the previous book of *Society 5.0*, we argued that the Society 5.0 vision encompasses a broader range of sectors compared to Germany's Industrie 4.0. It includes smart cities (urban or rural areas optimized for living and working with the implementation of digital solutions), industry (factories and workshops), energy, and other sectors, with a particular focus on smart cities. In Chap. 3, we explore six key factors pertinent to the Society 5.0 architecture, emphasizing the necessity and importance of each.

The Cabinet Office's Society 5.0 Reference Architecture (Cabinet Office 2020) is notable for its emphasis on data interoperability. However, this reference architecture is quite generic and, in our view, requires augmentation to effectively support the creation of a people-centric, sustainable smart city within actual urban or community settings. We propose three complementary perspectives—interface, organization, and process—which are extensively discussed in Chap. 3. These perspectives are further divided into six key factors, which are essential for any city or community. Throughout this book, we detail the significance of each factor and describe their practical application based on the findings from H-UTokyo Lab.

Cyberspace as a New Frontier in Urban Planning

In urban planning, new frontiers have opened whenever new technologies were developed and new concepts of living spaces were introduced in accordance with the demands of the times.

During the late nineteenth and early twentieth centuries, the time when the Briton Ebenezer Howard was active (see Prologue), suburbia was the new frontier in urban planning. Howard advocated turning suburban areas into garden cities that combined the benefits of the countryside and town. He also advocated the use of business management principles in the development of the garden cities. In this way, he pioneered a new kind of living space.

During the postwar years, Japan experienced a surge in bay reclamations, leading to the expansion of coastal industrial belts. Subsequently, many waterfronts became the focus of major redevelopment projects, which spurred a boom in more comfortable and spacious housing designs. Urban planners then shifted their focus upward: during the Japanese asset price bubble of the late 1980s and early 1990s, general contractors explored the concept of constructing ultra-tall skyscrapers, over 1 km in height, and developing the necessary technology. This initiated a competitive race to design ever taller skyscrapers in what became known as the "skyfront" era. Later, urban planners looked downward, developing infrastructure to create "geo-fronts" more than 40 m underground. With the growing interest in aerospace development, we have entered the age of space-fronts—an era in which ordinary members of the public can participate in spaceflight. Against this backdrop, the emergence of smart cities, which leverage cyberspace, marks a new phase in urban planning where cyberspace represents the latest frontier.

This new frontier is thrilling, yet it encompasses numerous risks, including those associated with cybersecurity, social ethics, and overall security. If we are to apply the same developmental processes used for previous physical frontiers to this new digital frontier and aim to make it a safe space supported by advanced technological innovation, then both sociological and engineering technologies will play critical roles. Moreover, building smart cities as integral components of the people-centric, super-smart society—which aims to seamlessly integrate cyber and physical spaces—requires that cyberspace be developed through secure methods. These methods must incorporate both the technologies and theories of the social sciences, as well as engineering technology. For these reasons, it is crucial to utilize the defined key factors and their maturity levels.

Applying the Key Factors as Infrastructure for a Digital Society

We have detailed key factors in Chaps. 4, 5, 6, 7, 8, 9, 10 and 14, which can be regarded as strategies to enhance the Cabinet Office's Society 5.0 Reference Architecture (Cabinet Office 2020) and the smart-city reference architecture (Cabinet Office 2023). These strategies are aimed at elaborating and practicalizing the reference architectures with the assumption that they will be implemented in real-world cities and communities. These key factors are intended to constitute the social infrastructure necessary for executing smart-city projects as businesses and services within the target city or community. Additionally, they are designed to ensure that smart-city projects are sustainable and people-centric, focusing on the well-being of residents, rather than being merely ephemeral trends.

In Chaps. 11, 12, and 13, we explored models for applying the Society 5.0 architecture in the context of the opportunities that arise when utilizing the reference architecture and key factors across three thematic areas. In the first instance, we demonstrated the potential of these frameworks to support citizen participation in urban planning through a data-driven system that visually represents foot traffic data and stimulates public dialogue (Chap. 11). In the second instance, we showcased their applicability in supporting smart aging through frailty-prevention services (Chap. 12). In the third instance, we highlighted their utility in managing urban infrastructure, such as roads and sewage systems (Chap. 13). In each scenario, we presented the implications of our findings—the challenges identified and the potential benefits.

Throughout this book, we have illustrated how the six key factors align with the Society 5.0 philosophy and underscored the necessity for every smart city to incorporate these elements. These factors essentially constitute the digital infrastructure required to properly develop the new frontier of cyberspace.

In conclusion, we anticipate that these six key factors, along with their defined maturity levels, will be effectively utilized as the digital infrastructure for developing new, secure, and trusted sectors within cyberspace.

Developing Human Resource in Accordance with the Society 5.0 Architecture

Finally, we would like to emphasize the critical importance of human resource development, as discussed in Chap. 9 of this book, for the realization of Society 5.0. The reference architecture of Society 5.0 consists of eight hierarchical layers. Among these, the upper layers—namely, "Strategy and Policy", "Rules" and "Organization"—are highly dependent on the capabilities of administrative bodies and public officials who are expected to take responsibility for their implementation. The middle layers, which include "Business", "Functions", "Data", and "Data linkage", require the expertise and commitment of private-sector actors. These include businesses, vendors, and other professional organizations that are expected to play a central role in their development and execution. The eighth and final layer, known as "Assets", is particularly reliant on effective collaboration between the public and private sectors. Public-private partnership plays a decisive role in the success of this layer. To optimize outcomes across all these layers, it is essential to develop expert teams that are well-versed in the Society 5.0 frameworks.

In particular, the creation of people-centric and sustainable smart cities cannot be accomplished by government agencies or private companies alone. It is essential to promote collaboration among public institutions, private enterprises, and academic and research institutions such as universities. In addition, as mentioned in Chap. 7, it is also necessary to advance these initiatives with thoughtful consideration for citizen engagement. For that purpose, fostering digital literacy among citizens and residents is vital. It is important that people develop the ability to properly understand and make use of digital technologies and to interpret data accurately.

We sincerely hope that the contents of this book will contribute to the development of professionals who will lead smart city initiatives based on the concept of the Society 5.0 architecture. We also hope that it will serve as a helpful resource for deepening the public's understanding of Society 5.0.

References

Cabinet Office (2020) Smart City Reference Architecture White Paper, 1st edn. In: Cross-ministerial Strategic Innovation Promotion Program (SIP) Second Phase, Big-data and AI-enabled Cyberspace Technologies/Smart City Architecture Development/Smart City Architecture Design and Promotion of Related

Verification Research (Released on March 31, 2020). https://www8.cao.go.jp/cstp/stmain/20200318siparchitecture.html. Accessed on March 7, 2024.

Cabinet Office (2023) *Sumāto Shiti: Rifarensu ākitekucha howaito pēp*ā [Smart city: Reference architecture white paper], 2nd edn. In: Cross-ministerial Strategic Innovation Promotion Program (SIP) Second Phase, Big-data and AI-enabled Cyberspace Technologies/Smart City Architecture Development /Smart City Architecture Design and Promotion of Related Verification Research (Released on August 10, 2023). https://www8.cao.go.jp/cstp/stmain/20230810smartcity.html. Accessed March 7, 2024.

H-UTokyo Lab (2020) Society 5.0: A People-centric Super-smart Society. Springer, Singapore. https://www.springer.com/gp/book/9789811529887. Accessed April 30, 2024.

Glossary

Society 5.0 (See Prologue) Society 5.0 is a vision for a new society set out in the Government of Japan's 5th Science and Technology Basic Plan, approved by the Cabinet on January 22, 2016. The vision is named "Society 5.0" because the envisaged society will be the "fifth" developmental stage of human society created by transformations led by scientific and technological innovation, following the hunter-gatherer (1.0), agricultural (2.0), industrial (3.0), and information (4.0) societies (Cabinet Office 2016a). In the government 2016 (Cabinet Office 2016b) and 2017 (Cabinet Office 2017) Comprehensive Strategy on Science, Technology, and Innovation (STI), Society 5.0 is defined as "a human-centered society that, through the high degree of merging between cyberspace and physical space, will be able to balance economic advancement with the resolution of social problems" "to ensure that all citizens can lead high-quality, lives full of comfort and vitality (Cabinet Office 2016b)."

Similarly, in the government's sixth 5-year strategy for science, technology, and innovation (approved by Cabinet in March 2021), Society 5.0 is defined as a "universal and global image of our future society." It further defines this society as one that is "sustainable and resilient, that secures the safety and security of the people, and that enables each and every one of them to realize diverse happiness (well-being)." (Cabinet Office 2021).

Architecture (See Prologue, Chap. 1) Architecture (as the Japanese loanword *ākitekucha*) is originally an architectural term referring to architectural techniques and designs or to architectural styles that are associated with a particular place or time. In this context, architecture can connote a method, technique, or theoretical model for designing and building architectural spaces that reconcile the abstract requirements of the client (living space requirements, work requirements, recreational needs) with the physical constraints of the locale, such as the local geographic and climatic features, the available materials (stone, timber, soil, and so on), and local lifestyles. Architecture can also refer to the physical buildings that are the products of architecture.

© Hitachi-UTokyo Laboratory (H-UTokyo Lab.) 2025
Hitachi-UTokyo Laboratory (H-UTokyo Lab.), *The Architecture of "Society 5.0"*,
https://doi.org/10.1007/978-981-96-2929-9

As a conceptual model, architecture can describe, in addition to real-world buildings and cities, the digital "architecture" for information systems that operate cyberspace, including data, software, and hardware.

In this book, we refer to Society 5.0 architecture. Society 5.0 architecture is, in addition to real-world urban structures, the architecture for the information systems in cyberspace. It also includes the architecture for social and industrial systems.

Reference Architecture (See Sect. 1.4) A reference architecture is a model and protocol describing the recommended way to deliver a technology. Its recommendations are nonmandatory and non-prescriptive (the designer chooses whether or not to adopt the recommended structures and designs); but by clarifying the set of recommended procedures, it helps ensure that all the requirements are met (that no required work is omitted), and by presenting a common, standardized guide and nomenclature, it helps avoid arbitrary initiatives. The Government of Japan has provided a reference architecture for Society 5.0 along with a smart-city architecture based on that. The government defines reference architecture as something that is to be referred by those who intend to implement Smart City, in order to confirm the necessary components required for the implementation and the relationships between them as well as required relationships with outside the Smart City (Cabinet Office 2023).

Smart City (See Chap. 2) A smart city encompasses the application of ICTs to enhance various aspects of urban management, including planning, development, management/operation (*means*). It represents an ongoing process aimed at addressing the needs and challenges of a city or community while fostering innovation and creating new value (*action*). Furthermore, it signifies a state of being a sustainable city/region where Society 5.0 is realized ahead of the others (*state*) (Cabinet Office et al. 2021). Figure A.1 illustrates the definitions of smart cities provided by the Government of Japan (represented by MLIT and the Cabinet Office), as well as those by international entities such as the UN (represented by the International Telecommunication Union), EU, and OECD. In this manuscript, we define a smart city as something that professes and embodies the people-centric ethos of Society 5.0.

Super City (See Sect. 2.3) "Super city" is a designation for a type of national strategic special zone in Japan, formally "Super-City National Strategic Special Zone." Such zones are supposed to pioneer the participatory, resident-focused initiatives of the future society Japan wants to achieve by 2030. To gain super city status, the city must have ideas for delivering innovative services that span different sectors and encompass all aspects of everyday life, linking data across different sectors, and far-reaching regulatory reforms (Secretariat for Promotion of Regional Revitalization, Cabinet Office 2024). In September 2020, Japan revised the National Strategic Special Zone Act, paving the way for cities' applications for super city status. Applications began in December 2020. In April 2022, the Cabinet awarded super city status to the cities of Osaka and Tsukuba.

		A concrete noun (a place) A means **An end**
	Definer (year)	Smart-city definition
Goverment of Japan's definitions	Smart-City Public– Private Partnership Platform (2019)	A project that uses ICTs to upgrade and streamline urban and community functions and services **to address issues and needs** and **create new value** that includes comfort and convenience; **a place where Society 5.0 is realized ahead of the others.** (MLIT 2024a)
	MLIT City Bureau (2019) Cabinet Office (2020)	A **holistically optimized**, sustainable **city or district** where management (planning, building-up, management/operation, etc.) is executed leveraging such advanced technologies as ICT for the resolutions of various issues of the city. (Cabinet Office 2023)(MLIT 2018)(MLIT2024b)
	Cabinet Office (2021)	A sustainable **city or region** that **solves challenges** faced by cities and regions, and continues to **create new value**, by providing services to support each one of residents using new technologies, such as ICT, and various public and private data, and by enhancing management in various fields (e.g. planning, development, management / operation) on the basis of the three basic philosophies and five basic principles; **a place where Society 5.0 is realized ahead of the others.** (Cabinet Office et al. 2021b) (Cabinet Office et al. 2023b)
	Cabinet Office (2021)	Sustainable **cities or regions** that utilize new technologies such as ICT while **resolving various issues** faced by the city or region through the sophistication of management (planning, development, control, operation, etc.) and continue to create **new value**). (Cabinet Office 2021a, p.16)

Fig. A.1 What is a smart city? A comparison of international definitions"Smart city" has been defined by Government of Japan (as represented by MLIT and the Cabinet Office) and by international bodies: the UN (as represented by the ITU), EU, and OECD

Super-Smart Society (See Prologue) A super-smart society is a future society envisaged in the fifth Science and Technology Basic Plan (approved by Cabinet in January 2016) (Cabinet Office 2016a). The government defines such a society as one "that will bring wealth to the people, through an initiative merging the physical space (real world) and cyberspace by leveraging ICT to its fullest."

Data-Linkage Platform (See Chap. 10) A data-linkage platform is a platform for collecting, collating, and sharing datasets across multiple sectors (Secretariat for Promotion of Regional Revitalization 2019). In the context of a super city, it describes a platform that is constructed using a building-block model and uses a

Definitions by international organizations	ITU (2016)	A smart sustainable **city** is an innovative city that ICTs and other means to improve QoL, efficiency of urban operation and services, and competitiveness, while ensuring that it meets the needs of present and future generations with respect to economic, social, environmental as well as cultural aspects. (ITU 2024) Note: "City Competitiveness" refers to the policies, frameworks, strategies, and processes determining the city's sustainable productivity.
	EU etc. (2017)	A smart city is a **city** that efficiently mobilizes and uses available resources (including but not limited to social and cultural capital, financial capital, natural resources, information and technology) for efficiently • **improving the QoL** of its inhabitants, commuting workers and students, and other visitors [people] • significantly improving its resource efficiency, decreasing its pressure on the environment and **increasing resiliency** [planet] • **building an innovation-driven and green economy** [prosperity] • **fostering a well-developed local democracy** [governance] (EU et al. 2017)
	OECD (2020)	What is central to the smart-city definition is how digitalization helps achieve four core objectives, i.e., **improve people's well-being** and **foster more inclusive, sustainable, and resilient societies**. (OECD 2020)

Fig. A.1 (continued)

public application programming interface to accumulate and distribute data. The platform consolidates datasets provided by various parties, converts the data into the necessary formats, and then distributes the data using an application programming interface (Secretariat for Promotion of Regional Revitalization 2020).

Urban Operating System (See the "Column" in Chap. 10) Urban operating system is an umbrella term for IT systems used by communities that are trying to create a smart city. It is used to consolidate the functions that the community will use interoperably and to facilitate the delivery of multi-sectoral services in the smart city (Cabinet Office 2023). An urban operating system enables utilization of any combinations of services and functions provided by various operators and other regions (Cabinet Office 2020).

In a smart city, an urban operating system has data-linkage functions (data management, linkage with external data, asset management) and functions related to service linkage, authentication, and service management.

Plateau (See Sect. 2.5) Plateau is a 3D city modeling program, an example of the digital infrastructure that will drive the digital transformation in smart-city projects and other community-building projects. It has been developed and applied with its findings published as open data, as part of a MLIT project. https://www.mlit.go.jp/plateau/about/

Six Key Factors (See Chap. 4) H-UTokyo Lab proposed the six key factors as an approach for delivering a smart city that embodies the people-centric ethos of Society 5.0. They include factors for building a smart city together with the community (citizen participation [in a living lab], social acceptance), those related to people-centric data use (QoL-based assessment, data governance), and those related to ensuring that the smart city is sustainable (data ecosystem, human resource development).

Social Acceptance (in a Smart City) (See Chap. 5) Social acceptance refers to a situation in which the community is well-informed about the smart-city initiatives, actively participates in the decision-making process, and welcomes the initiatives. In our view, social acceptance is not about satisfying a generic set of legal or regulatory criteria to use personal data. It must involve a case-specific approach for building trust among the community stakeholders.

Data Governance (in a Smart City) (See Chap. 6) Data governance refers to a situation in which smart-city data, including personal data and impersonal data (data related to city operations), is governed in accordance with a well-defined set of rules. This means more than complying with personal data legislation such as those set out in Japan's Act on the Protection of Personal Information; it also means governing the data in a way that will avert controversies in which residents feel that their privacy is threatened, imperiling the continuity of the service.

Living Lab (Citizen Participation) (See Chap. 7) A living lab is a series of processes, a place, or a technique in which members of the public and other stakeholders come together to engage in co-creation projects involving pilots and trials conducted in real-world environments. It can refer to the following series of processes as a whole or to one or some of these processes: an exploratory analysis to identify the community's needs and issues, forming a vision, designing ideas, and developing services. To create a smart city, the members of the public who engage in co-creation with other stakeholders should be those who know the community's needs and issues the most, and their knowledge, experience, and creativity should be leveraged. A living lab is a co-creation space that satisfies this requirement.

Smart-City QoL-Based Assessment (See Chap. 8) A smart-city QoL-based assessment is a new approach to assessing the outcomes of smart-city programs, one that uses people-centric measures such as well-being and quality of life. The approach captures the satisfaction and well-being of residents of the smart city, in terms of the extent to which they are living meaningful and authentic lives and experiencing happiness.

Well-Being (See Chaps. 2 and 8) The definition of well-being is found in the preamble to the WHO charter: "Health is a state of complete physical, mental and social well-being and not merely the absence of disease or infirmity (WHO 1946)." It has also been defined as degree of satisfaction (Cabinet Office 2024). In the definition of Society 5.0 provided in the government's sixth 5-year strategy for science, technology, and innovation, well-being is equated with diverse forms of happiness: "(Society 5.0 is a society that) is sustainable and resilient, that provides safety and security for citizens, and that secures the safety and security of the people, and that enables each and every one of them to realize diverse happiness (well-being). (Cabinet Office 2021)".

Data Ecosystem (See Chap. 10) A data ecosystem is an economic ecosystem mediated largely by data formed through the proliferation of cross-sectional data use. It comprises networks of data providers and data users and it is a set of social connections through which companies and other organizations share data with each other and through which valuable services are delivered and the value of such is consumed.

Data-Driven Urban Planning (See Chap. 11) Data-driven urban planning occurs when decisions about new urban interventions are made based on a comprehensive analysis of various urban data sources. This process utilizes a wide spectrum of data—including foot traffic, road traffic, congestion, environmental, energy, and purchasing data—collected from payment services to simulate potential outcomes of proposed urban interventions. The results of these simulations provide an evidential basis for evaluating and making informed decisions regarding these interventions.

Smart Aging Society (See Chap. 12) A smart aging society is one that adapts to the challenges posed by a super-aging population. It promotes independence and community engagement among elderly people, enabling them to remain active throughout their lives. This society is characterized by its inclusiveness and commitment to co-creation among all generations.

Frailty (See Sect. 12.1) Frailty, adopted as the Japanese loanword *fureiru*, is a concept introduced by the Japan Geriatrics Society in 2014. Based on the English term "frailty," it describes an aging-related decline in physical and mental faculties that diminishes the person's ability to withstand, or recover from, stressors.

Value-Creating Infrastructure (See Chap. 13) Value-creating infrastructure refers to a process for the sustainable management of community infrastructure. This approach extends beyond mere maintenance of built infrastructure; it anticipates long-term changes within the city or community and identifies challenges that may arise in the future. The primary objectives of this process are twofold: (1) to enhance the infrastructure's value to residents, and (2) to maintain and renew the infrastructure as necessary.

References

Cabinet Office (Council for Science, Technology and Innovation) (2016a) *Dai 5 ki kagaku gijutsu kihon keikaku* [The 5th Science and Technology Basic Plan] (Released on January 22, 2016). https://www8.cao.go.jp/cstp/kihonkeikaku/index5.html. Accessed March 28, 2024.

Cabinet Office (Council for Science, Technology and Innovation) (2016b) *Kagaku gijutsu inobēshon sōgō senryaku 2016* [Comprehensive Strategy on Science, Technology, and Innovation 2016] (Released on May 24, 2016). http://www8.cao.go.jp/cstp/sogosenryaku/2016.html. Accessed May 30, 2023.

Cabinet Office (Council for Science, Technology and Innovation) (2017) *Kagaku gijutsu inobēshon sōgō senryaku 2017* [Comprehensive Strategy on Science, Technology, and Innovation 2017] (Released on June 2, 2017). http://www8.cao.go.jp/cstp/sogosenryaku/2017.html. Accessed May 30, 2023.

Cabinet Office (2020) How to Use Smart City Reference Architecture: Installation Guidebook, English Edn. (Released on August 1, 2020) https://www8.cao.go.jp/cstp/stmain/a-guidebook1_200331_en.pdf. Accessed October 23, 2024.

Cabinet Office (Council for Science, Technology and Innovation) (2021) The 6th Science, Technology, and Innovation Basic Plan (Released on March 26, 2021). https://www8.cao.go.jp/cstp/english/sti_basic_plan.pdf. Accessed May 7, 2024.

Cabinet Office (2023) *Sumāto Shiti: Rifarensu ākitekucha howaito pēpā* [Smart city: Reference architecture white paper], 2nd edn. In: Cross-ministerial Strategic Innovation Promotion Program (SIP) Second Phase, Big-data and AI-enabled Cyberspace Technologies/Smart City Architecture Development /Smart City Architecture Design and Promotion of Related Verification Research (Released on August 10, 2023). https://www8.cao.go.jp/cstp/stmain/20230810smartcity.html. Accessed March 7, 2024.

Cabinet Office (2024) *Well-being ni kansuru torikumi: Manzokudo seikatsu no shitsu ni kansuru chōsa* [Well-being Initiatives: Survey on Satisfaction and Quality of Life]. https://www5.cao.go.jp/keizai2/wellbeing/index.html. Accessed October 24, 2024.

Cabinet Office, Ministry of Internal Affairs and Communications, Ministry of Economy, Trade and Industry, Ministry of Land, Infrastructure, Transport and Tourism and Smart City Public-Private Partnership Platform Secretariat (2021) Smart city guidebook, ver 1.01, April, 2021. https://www8.cao.go.jp/cstp/society5_0/smartcity/01_scguide_eng_1.pdf. Accessed October 25, 2024. https://www8.cao.go.jp/cstp/society5_0/smartcity/olddocument.html. Accessed March 28, 2024.

Cabinet Office, Ministry of Internal Affairs and Communications, Ministry of Economy, Trade and Industry, Ministry of Land, Infrastructure, Transport and Tourism and Smart City Public-Private Partnership Platform Secretariat (2023)*Sumātoshiti gaidobukku* [Smart city guidebook], 2nd Edn., April 2023. https://www8.cao.go.jp/cstp/society5_0/smartcity/guidebook.html Accessed October 25, 2024.

EU, Bosch P, Jongeneel S, Rovers V, Neumann HM (2017) CITYkeys Indicators for Smart City Projects and Smart Cities. https://www.researchgate.net/publication/326266723_CITYkeys_indicators_for_smart_city_projects_and_smart_cities. p.7 Accessed October 25, 2024.

ITU (2024) ITU-T Y.4900/L.1600 recommendations. https://www.itu.int/itu-t/recommendations/rec.aspx?rec=12627 Accessed October 25, 2024.

Ministry of Land, Infrastructure, Transport and Tourism (MLIT) (2018) *Sumātoshiti no jitsugen ni mukete: Chūkan torimatome* [Toward the Delivery of Smart Cities: Interim Report], April 2018. https://www.mlit.go.jp/common/001249774.pdf. Accessed October 25, 2024.

MLIT (2024a) Smart-City Public–Private Partnership Platform. https://www.mlit.go.jp/scpf/index.html#home02. Accessed August 18, 2023.

MLIT (2024b) *Sumātoshiti ni kansuru torikumi* [Smart-City Programs]. https://www.mlit.go.jp/toshi/tosiko/toshi_tosiko_tk_000040.html. Accessed August 18, 2023.

OECD (2020) Measuring smart cities' performance. https://opencommons.org/images/c/c1/Smart-cities-measurement-framework-scoping.pdf. p.13 Accessed October 25, 2024.

Secretariat for Promotion of Regional Revitalization, Cabinet Office (2019) *Sūpā shiti to dēta renkei kiban ni tsuite* [About super cities and data-linkage platforms] (materials used at the 39th meeting of Council on National Strategic Special Zones, April 19, 2019). https://www.chisou.go.jp/tiiki/kokusentoc/dai39/shiryou3_2.pdf. Accessed October 23, 2024.

Secretariat for Promotion of Regional Revitalization, Cabinet Office (2020) *Sūpā shiti/sumāto shiti no sōgō unyō sei no kakuho tō ni kansuru kentōkai saishū hōkokusho* [Final report of the committee for ensuring data interoperability in super cities and smart cities], September 2020. https://www.chisou.go.jp/tiiki/kokusentoc/supercity/pdf/sogowg_houkokusyo.pdf. Accessed October 23, 2024.

Secretariat for Promotion of Regional Revitalization, Cabinet Office (2024) *Sūpā shiti dijitaru den'en kenkō tokku ni tsuite* [About super cities and digital garden health special zones]. January 2024. https://www.chisou.go.jp/tiiki/kokusentoc/supercity/supercity.pdf. Accessed March 15, 2024.

Wordl Health Organization (WHO) (1946) Constitution of the World Health Organization. https://www.who.int/about/governance/constitution. Accessed October 24, 2024.